CHICAGO'S WAR ON SYPHILIS, 1937–1940

Chicago's War on Syphilis, 1937–1940

The Times, the *Trib*, and the Clap Doctor

With an epilogue
on issues and attitudes
in the time of AIDS

SUZANNE POIRIER

University of Illinois Press • Urbana, Chicago, and Springfield

Manufactured in the United States of America

This book is printed on acid-free paper.

Library of Congress Cataloging-in-Publication Data
Poirier, Suzanne.
Chicago's war on syphilis, 1937–1940 : the Times, the Trib, and the Clap Doctor / Suzanne Poirier.
p. cm.
Includes index.
ISBN 978-0-252-02147-3 (cloth)
1. Syphilis—Illinois—Chicago—History—20th century.
2. Reitman, Ben L. (Ben Lewis), 1879–1942. 3. Chicago Tribune.
4. Chicago Sun-Times. I. Title.
[DNLM: 1. Reitman, Ben L. (Ben Lewis), 1879–1942. 2. Syphilis—prevention & control—Chicago. 3. Syphilis—history—Chicago.
4. Public Health—history—Chicago. WC 160 P753C 1995]
RA644.V4P65 1995
362.1'969513'00977311–dc20
DNLM/DLC
for Library of Congress 94-19357
CIP

To my parents, Ruthe and Kenneth Poirier,
and to Mecca Reitman Carpenter, for sharing her parents with me.

Contents

Acknowledgments

This book owes its existence to the magnanimity, advice, and encouragement of many people. It began and, most appropriately, concluded with Ruth and Joel Surgal, possessors of that proverbial "box of papers in the attic." Even more important than living in a house once owned by the son of "The Clap Doctor of Chicago," they knew Ben Reitman's history and the value of historical documents. Ruth brought me the first box, "to see if there's anything here," and started me on a journey that has proved intriguing, challenging, and always entertaining. As my work was, I thought, winding to a close, she found a last cache of letters and photographs in another trunk that filled in last gaps, confirmed some hunches, and offered a wealth of photographs not previously archived. Certainly, the story that is told here would never have emerged without the Surgals' curiosity and generosity.

Equally generous has been Mecca Reitman Carpenter, Ben Reitman's eldest daughter from his union with Medina Oliver. My research has introduced me to many delightful people, but Mecca has become a friend. When I met her, Mecca was already working on her own book, a biography of her father and mother—work that demands an encyclopedic familiarity with the extensive papers of a number of people as well as the devotion and stamina to work so closely on the personal lives of people with whom her own life is so entwined. In sharing her own research with me, she has also shared parts of her life, a rare and precious collaboration.

This is not the usual book that a literature scholar would write, and this untraditional—or rash—project owes much to the support and wisdom of several historians of medicine with whom I have had the privilege to work.

Dan Jones, through his own love of the process of historical research, taught me how to think like a historian; Patricia Spain Ward, through the example of her joyous energy for research, taught me tenacity and persistent curiosity; and Norman Gevitz, through his steadfast confidence in me (coupled with tactful guidance and tutoring), often kept me from jumping ship or settling for easy answers.

Mary Ann Bamberger, assistant special collections librarian at the University of Illinois at Chicago, not only helped me find my way through Ben Reitman's sizable collection but also introduced me to numerous people, including Mecca Carpenter, whose work has been useful to me. Terry Moon and Zita Stukas helped me time and again work efficiently with this large collection. Roger Bruns, Allan Brandt, and Thomas Bonner read parts of my manuscript and directed me to further, invaluable resources; Patricia Spain Ward has been a reference library unto herself; and Tim Murphy has been a tireless and—fortunately for me—remorseless reader and commentator. An invitation to work for six months with the Medical Humanities faculty at East Carolina University provided a valuable time and space for rethinking and revision. I owe that department's director, Loretta Kopelman, a special thanks for this opportunity.

At the University of Illinois Press, Judith McCulloh receives a special thanks for having faith in and patient support for a project whose evolution was, at least initially, slow and murky. More in the trenches, I am equally indebted to Mary Giles, who brought this book into its final form.

INTRODUCTION

Of Snakes and Snake-Killers

"Drag the snake out of the bushes and beat its head off in public."[1] The order came from Herman N. Bundesen, commissioner of the Chicago Board of Health. The snake was syphilis. The year was 1922. That snake, as is the nature of snakes, proved to be hard to catch and even harder to kill. However much it might have been beaten in public, its head stayed on.

Fifteen years later syphilis reappeared, transformed into fire but still evoking evil and danger. Syphilis had "been allowed to smoulder in our midst all too long," charged Irving S. Cutter, M.D., health editor of the *Chicago Tribune*.[2] Although the menace of syphilis had not abated, the tone of the metaphor had shifted. Where in 1922 the public was rallied to action, by 1937 syphilis, "allowed to smoulder," seemed to have an upper hand—moreover, to have been ceded that power by a passive public. Although that public had once aggressively hunted snakes, it was now ignoring a threatened blaze—and in Chicago at that! What had become of the snake hunters of 1922? How does a snake differ from a smoldering fire, and is one easier to fight than the other?

Dr. Cutter's metaphor was inspired by the creation of the Chicago Syphilis Control Program, one part of a national initiative to bring all cases of syphilis under treatment. The campaign depended upon public support in word and deed for its success as much as it did upon funding and personnel. Although it was not the only city involved in a program of syphilis control, Chicago was specifically selected to be a flagship city for this national undertaking.[3] Thus, while a part of Chicago's own unique history, the Chicago Syphilis Control Program also presents a microcosm of the treatment and control of venereal disease in the United States in the late 1930s.

In its fight against syphilis, the Chicago Syphilis Control Program used language as well as medicine, persuasion as well as science. The following study, therefore, weighs image and argument equally with policy and practice. Neither words nor acts, however, exist in a vacuum.[4] Although public interest and opinion—seasonal, fickle, and not always rational—often proceed without a sense of either past or future, they are inseparable from their own cultural past and present. The chapters that follow explore the many contexts within which the program evolved and how those contexts contributed, separately and collectively, to the program's successes and limitations.

Syphilis before 1937

The story of the Chicago Syphilis Control Program embraces language, law, medicine, morals, people, and places. It demonstrates that the complex interrelationships of individuals and circumstances can abet or confound any organization's "best laid plans." The main protagonist (or antagonist) of this story, whether snake or smoldering fire, is syphilis, with gonorrhea playing a strong but often unremarked supporting role.[5] Both diseases have a long history. Galen named gonorrhea in 130 C.E., the still popularly used *clap* first appearing in print in 1378. Hippocrates described genital sores that were probably primary syphilitic lesions in fourth century B.C.E., but syphilis did not appear in European medical records until the mid-1400s. For some time, experts were uncertain whether syphilis and gonorrhea were separate diseases or merely two manifestations of one disease, but in 1879 a German physician, Albert Ludwig Neisser, identified the gonococcus bacteria, accordingly named *Neisseria gonorrhoeae*. The corkscrew-shaped (hence *spirochete*) *Treponna pallidum* of syphilis, on the other hand, evaded scientists until 1905, because its narrow shape and high motility made it difficult to distinguish with the microscopes of that time.

Syphilis and gonorrhea were originally considered part of the same disease because of their sexual mode of transmission, but the courses and manifestations of the two diseases differ markedly. Syphilis is the more complex, protracted disease, identified by various characteristics at several identifiable stages: primary, secondary, and tertiary. Primary syphilis usually appears during the first two to four weeks following infection and is characterized by the presence of a nontender, firm, hard-edged ulcer in the genital or oral areas where inoculation has occurred. Called a chancre, this ulcer, if untreated, will heal spontaneously within ten to fourteen days. In women, chancres may form internally, and syphilis can go undetected in this primary stage. Most infections are transmitted during the primary stage, and about one-third of the people exposed to primary syphilis likewise become infected.

If not immediately treated, primary syphilis will progress to secondary syphilis, as the spirochete is disseminated throughout the body. Secondary syphilis usually manifests itself by extensive skin rashes; less common are sore throat, fever, or hepatitis. Some patients, however, may not exhibit these symptoms immediately but still test positive to syphilis. They are defined as having latent syphilis, a category further distinguished between early and late latency, the former designating infections thought to have been present for less than a year and the latter cases suspected to have been present for more than a year. (These definitions, from the Centers for Disease Control, are the ones in current use in the United States. The World Health Organization defines early and late latency in two-year periods. This lack of consensus reflects the lack of earlier testing in many infected people, who either did not suspect disease or did not seek testing for their symptoms.) Patients in early latency are potentially infectious (as are patients with secondary syphilis), and about a quarter of them go on to develop signs of secondary syphilis within the first year of their infection. Women who give birth during early latency may also infect their children.

Tertiary syphilis, for a long time not so identified, can occur years after the spontaneous clearing of secondary syphilis. It produces potentially fatal lesions of skin, bone, or tissue in the cardiovascular, central nervous, or musculoskeletal systems. About one-third of untreated patients develop tertiary syphilis. Thus, even if no longer infectious, untreated syphilis could kill the diseased person, a consequence whose full extent was not realized for many centuries, and asymptomatic patients were often mistakenly believed to be cured. Because of this ignorance about the extent and nature of tertiary syphilis, deaths from cardiovascular syphilis were often incorrectly attributed to heart disease, and neurosyphilis was misdiagnosed as insanity, meningitis, or paralysis. Syphilis' reputation as "the great imitator" grew from these diverse manifestations of both secondary and, especially, tertiary syphilis. Many of the problems with gathering epidemiological information about syphilis, as testified by the employees of the Chicago Syphilis Control Program, grew out of such diagnostic complexity. As scientists and physicians came to know more about tertiary syphilis, more people were suspected to have the disease. Consequently, the image of syphilis as a persistent, unremitting killer grew and developed its own mythology and moral interpretations.

Because of the difficulties of diagnosing syphilis in all three stages and the uncertainty about infectiousness and cure, much about syphilis remained unknown, even by the 1930s. Gonorrhea, on the other hand, was more fully understood, in part because of its shorter, more easily defined course. Its relative straightforwardness has served to make gonorrhea, in many people's minds, less dreadful than syphilis, although it has probably always been more widespread. Even today, with gonorrhea once more on the increase and, as always, more prevalent than syphilis, medical concern tends to focus simply on finding the next efficacious strain of penicillin to use against the disease. As one medical writer observes, "To many, gonorrhea has become boring. It is simply too familiar, too well understood, and too easily treated" to engage the imagination of medical researchers.[6]

Gonorrhea usually infects the genito-urinary tract (the urethra in men and the surface of the cervix in women), but it can infect the rectum, pharynx, and conjunctivae of the eye as well, where gonococci adhere to mucosal surfaces and enter the bloodstream. Initial infection in men is usually marked by pain on urination and a discharge, often copious and foul-smelling, from the penis one to eight days after exposure to the disease. If untreated it will resolve in a few weeks. In most cases, gonorrhea will become asymptomatic within six months. Women may exhibit painful urination and increased vaginal discharge or bleeding, but, as with syphilis, symptoms may go unnoticed because the site of infection is often internal. In both men and women, gonorrhea can become acute by spreading up the genito-urinary tract, causing pelvic inflammatory disease in women and severe infection in the urethra or inflammation of the epididymis (a structure in the testis) in men. Secondary complications from gonorrhea occur more frequently in women, with infertility and ectopic pregnancy the most common of these, but gonorrhea can sterilize men as well.

At first, diagnosis of gonorrhea and syphilis were similar procedures, requiring microscopic examination of secretions, or *smears*, from infected sites. Identification of *T. pallidum* of syphilis was more difficult because of its size and relative transparency (*pall-*, meaning *light*, attesting to the problem it posed its seekers). The addition of angled mirrors to the microscope, around 1906, enabled viewers to see the bacteria against a dark, rather than a light, background—thus the designation *dark-field examination*. By 1906, though, an easier method of diagnosing syphilis had been devised by August von Wassermann, Albert Neisser, and Carl Bruck. The Wassermann test, as it has generally come to be known, detects the presence of *T. pallidum* through its reaction to substances in the blood. A sample of blood is tested instead of a mucosal smear from an infected site. The test involves more time and sophisticated analysis than the immediate microscopic examination of a smear, but it has the advantage of not requiring the presence of the primary lesion, which, in the case of syphilis, permits diagnosis only during a very short time. It is also possible to obtain samples from large numbers of people with a minimum of space, personnel, and equipment, economies that made possible the creation of the "dragnet stations" in Chicago's city parks and clinics in the summer of 1938.

The difference between drawing blood and obtaining mucosal smears, then, made mass screening programs for syphilis possible. Still, the problem of inexpensive testing continued. In the early 1920s, another blood test was developed by physician Reuben L. Kahn (often referred to as a Wassermann as well, because the antibody that precipitated in Kahn's tests was the same one originally named for Wassermann). Kahn's procedure offered a cheaper but less reliable alternative to the original Wassermann. A two-step protocol was developed in which a positive result on the initial Kahn test led to a follow-up Wassermann test or, when possible, a dark-field examination for even greater accuracy. The danger of false positives, especially with the Kahn, was everpresent. Kahn himself wrote, in a personal letter, "But the physician who is in the habit of telling patients that they have syphili[s] because their blood tests are positive, is unworthy of the name physician."[7] (Today, a newer, even further refined version of the Kahn test remains the first screen for syphilis, but the continued problem of false positives requires a second, more sensitive test for verification.)

In the case of gonorrhea, blood tests to detect *N. gonorrhoeae* had not been devised by the time of the Chicago Syphilis Control Program. Although examination of a smear was reliable, the physical inconvenience of such an examination at citywide test sites made regular testing impossible. Early smears to test for gonorrhea were about equally as accurate as smear techniques for diagnosing syphilis, but it wasn't until the early 1960s that scientists developed the tests for gonorrhea that are in use today.

Because of the problems of diagnosis, under-reporting of syphilis and gonorrhea, and the unsophisticated methods of early epidemiologists, actual incidence and rate of infection for both diseases in the nineteenth and early twentieth centuries were uncertain. Estimates of percentage of the public infected with syphilis in the United States ranged from 10 to 25 percent, with estimates for gonorrhea usually running two to three times higher. Whatever numbers were offered, and however accurate the conclusion, venereal diseases were generally considered to be occurring in epidemic proportions in the early 1900s.

One reason why the United States Public Health Service (USPHS) launched its antisyphilis campaign in 1937 was a growing confidence in medicine's ability to treat syphilis. Even so, the treatment was lengthy and unremitting. It involved repeated dosages of various mixtures of heavy chemicals—one of the first uses of chemotherapy. Treatment was based on injections of arsenical compounds, first discovered by Paul Ehrlich in 1909 and named Salvarsan—or, popularly, 606, because Ehrlich's discovery occurred with his six-hundred-sixth experiment. Highly toxic and difficult to administer safely, Salvarsan was replaced in 1912 by Neosalvarsan, and its use steadily grew. The newer treatment still carried some risk, however. Moreover, it was lengthy and expensive. Treatment began with twelve injections each of neo-arsphenamine and bismuth twice a week, then eight neo-arsphenamine and sixteen bismuth injections given twice a week. A third course consisting of eight "neo" and twenty bismuth (or ten bismuth and ten mercury) injections might still be required. This treatment lasted nearly a year. After a month's rest, patients received a spinal test. Patients who tested negative would be permitted another month's rest and then, again weekly, given six "neo" and twenty bismuth injections, or some combination thereof, to conclude the treatment. If side effects occurred (and they often did), the treatment became even more lengthy and complicated. In all courses of chemotherapy, treatment was followed upon completion with quarterly blood tests for one year and biennial tests for the next two.[8]

In Chicago in 1937, the average cost of treatment, at two visits a week for the first two years, could range from $150 to $500 a year, depending on whether the patient went to a private physician or saw a physician in one of the large, low-cost clinics.[9] Both money and motivation were required to complete a course of treatment, especially if the disease did not respond to the drugs as "quickly" as two years. Among medical and scientific experts, however, there was no clear agreement as to when a cure had actually been effected. In fact, many people with syphilis were probably subjected to this debilitating treatment longer than really necessary, because once the dangers of false positives and the persistence of syphilis were realized, overtreatment became as likely as undertreatment.[10]

Obstacles to treating gonorrhea were even greater than those to treating syphilis. Although by 1938 there was small but vocal support for the new "miracle" drug, sulfanilamide,[11] its ability to cure gonorrhea was still being debated. In the years before and during the Chicago Syphilis Control Program, the Public Health Service was researching new tests and efficacious treatments for gonorrhea, and their annual reports as well as those of the program contained statistics on gonorrhea. In their public rhetoric, however, although they prophesied the eradication of syphilis, they remained noticeably silent about their prospects against gonorrhea.

Finally, if medical and physical complexities weren't enough to complicate the lives of people with syphilis and gonorrhea, the moral and social climate around the diseases made them even harder to bear. Even before the Public Health Service became deeply involved in fighting venereal disease, professional and public disagreement raged over whether syphilis was a medical or a moral problem. The very term *venereal disease* demonstrates the tension within that debate. Carrying the definition "sexual indulgence," *venereal* implies moral condemnation from the outset.[12] From their ages-old association with illicit sex, syphilis and gonorrhea were often regarded as proof of moral laxness in the men and women infected with these diseases. Because intimations of immorality and sexual lasciviousness were also often part of particular racial or class stereotypes, "venereal disease" was often associated with specific—usually socially oppressed or disadvantaged—groups of people as well. Clearly, it is the human dimension of syphilis and gonorrhea—of all sexually transmitted diseases—that has made detection, treatment, prevention, and cure so elusive. It was with this dimension of syphilis that the Chicago Syphilis Control Program was to have the least success.

Creation of the Chicago Syphilis Control Program

Although the Chicago Syphilis Control Program made its public debut on January 18, 1937, its carefully orchestrated inauguration was preceded by months of careful preparation and negotiation on the federal, state, and city levels. "The venereal diseases are beginning to receive the attention in public health work that their relative importance justifies," Surgeon General Thomas Parran reported to the United States Congress in July 1936.[13] Parran was referring specifically to money recently made available to states through the Social Security Act, which would help the Health Service make improvements in state, city, and private laboratories' ability to perform accurate, efficient tests for syphilis and gonorrhea. These funds also provided states with special advisors to help them design and implement their own venereal disease control programs. During the next fiscal year, the Public Health Service sent advisors to seventeen states.[14]

Illinois was an early recipient of these Social Security funds. The first allocation occurred in June 1936, after Surgeon General Parran himself visited with Health Commissioner Bundesen about the board's present and projected work in syphilis control. Bundesen submitted a budget request to Frank J. Jirka, director of Public Health for Illinois, for a share of the state's Social Security funds. He copied the request to Parran, with a letter stating that "the Chicago Board of Health [was] the logical agency to undertake" initiatives of interest to the Public Health Service in the Chicago area.[15] Bundesen's bid for a leading role in the new programs, in effect supplanting the role usually assumed by state departments of health, signals tensions between Chicago and the state office.[16] The letter also indicates, however, that the Health Service focussed its attention on Chicago early in its national campaign.

New plans and old tensions were both more clearly defined in early December, after officials of the Public Health Service again visited Springfield and Chicago. Most noteworthy were visits from Assistant Surgeon General Raymond A. Vonderlehr, head of the Division of Venereal Disease in the Health Service, and Dr. Oliver Clarence Wenger, the service's special consultant in syphilis control for the Chicago area. Both men had been recently assigned to these positions, taking them from Tuskegee, Alabama. There, Vonderlehr had efficiently overseen the Tuskegee Syphilis Study and Wenger had tirelessly built popular support for syphilis control among southern African Americans.[17] During their visit, state, local, and national officials discussed augmenting the money already allocated for fiscal year 1936 with an additional $94,000 of federal funds for "much needed epidemiologic work connected with the control of syphilis and for the purchase of a large supply of anti-syphilitic drugs." Most of the money would be used in Chicago, where Vonderlehr and Wenger met with Bundesen "to ascertain the plans which were being made for the control of syphilis in Chicago, [and] generally speaking, it met with [their] approval." Bundesen's office would "specialize in case finding and the medical follow-up of delinquent patients and . . . provide through his municipal clinic and through the clinic of the Illinois Social Hygiene League more efficient treatment for syphilis and gonorrhea."[18]

Although Vonderlehr approved the program, he expressed some concern about conflicts among the several groups and personalities upon whose cooperation the success of the program depended. "It is believed to be most unfortunate," Vonderlehr wrote Surgeon General Parran, "that Doctor Louis E. Schmidt still heads the venereal disease control work in the City of Chicago because of the animosity which exists between this physician and members of the Chicago Medical Society."[19] This "animosity" posed a problem. Vonderlehr needed the backing of the Chicago Medical Society for two surveys that would take place in the preliminary stages of the program. He expressed a desire that Schmidt's new assistant, Dr. George G. Taylor, be trained to take Schmidt's place. He also recommended that State Director Jirka be employed as an intermediary, because Jirka was "on unusually friendly terms with the Illinois Medical Society at the present time and his standing with them should place him in a favorable position to approach the Chicago Medical Society."

Vonderlehr concluded his letter to Parran with a further expression of concern: "Potentially a great deal should be accomplished in the control of syphilis both in Illinois and Chicago. Both health departments are very much interested in the control of this disease and desire to take up the campaign at once. The situation is disturbed because of the well known lack of cooperation between the State and City [H]ealth Departments and because of the animosity which exists between the City Health Department and organized medicine in Chicago." Whether or not a result of Vonderlehr's concerns, by the time the Chicago Syphilis Control Program was formally begun, Louis Schmidt's position had been changed from chief of the Division of Venereal Diseases to secretary of the Board of Health. George Taylor, newly promoted to division chief, was running the clinical operations of the Chicago Syphilis Control Program.

No program built on the shoulders of existing programs and departments will be without a history of old battles and ongoing irritations, so Vonderlehr's expressions of concern should not be read as foreshadowing inevitable disaster; in fact, the assistant surgeon general's perceptiveness from just one visit probably speaks more aptly to his own astuteness. Nevertheless, his comments signal the delicacy of some of the work that lay before them all.

By early December 1936, then, the U.S. Public Health Service had identified Chicago as the site for a specially concentrated effort toward syphilis control. The official kick-off of the *federal* campaign against syphilis occurred at a national Conference on Venereal Disease Control Work in Washington, D.C., on December 28–30, 1936. The surgeon general called the meeting "for the purpose of considering basic principles underlying a national plan for the control of venereal disease." President Franklin D.

Roosevelt sent his greeting to the more than nine hundred participants from forty-five states, Hawaii, Puerto Rico, and more than ten foreign countries. Publicity was as much a goal of the conference as were education and planning. A report of the conference concluded, "The increasing cooperation and widespread support of the venereal disease control program on the part of private physicians, voluntary agencies, and the public in general are doubtless due in considerable measure to the stimulus provided by this great national conference."[20]

Less than a month later, Mayor Edward J. Kelly of Chicago convened a meeting in his council chamber to announce the creation of the Chicago Syphilis Control Program, supported by money from the United States Public Health Service, Works Progress Administration, Illinois Department of Public Health, and Chicago Board of Health. It was hailed as "the first attempt to control syphilis in a large urban population by the combined efforts of Federal, State, and local agencies."[21] In a conference that did not exceed thirty-five minutes, members of Chicago's leading labor, philanthropic, medical, business, and religious communities voiced their support for the program, hailing its national as well as local significance. With the exceptions of Illinois Public Health Director Jirka and physician-activist Paul de Kruif, all of the speakers were Chicago men and women, most of whom formed the backbone of a larger group of fifty-nine leaders who had agreed to create subcommittees representing their various constituencies in the city.[22] The group would be called the Committee of 400 and would serve as advisors to and visible supporters of the Chicago Syphilis Control Program.[23] Everyone was determined to make the program a success.

The Perspectives

In January 1937, Chicago stood poised on the brink of a new public health initiative, and the city would be the laboratory for the nationwide undertaking. To many Chicagoans, though, it was a renewed effort, the third or fourth campaign in an ongoing, dauntless battle by a city that was not intimidated by convention or false modesty. And, to some individual Chicagoans, the Chicago Syphilis Control Program promised the fulfillment of years of dreams and hard work. The preceding summary of events reveals a complex web of policy and personality that reached back to the years before the program's creation and promised to influence its development from the start. Given this multiplicity of inputs and viewpoints, how do we discover, let alone understand, the work and achievements of this project?

Because this book looks at past events, some people would call its study history. Because it looks at the debate surrounding those events, other people would call it rhetorical analysis. But an account of the people and events of the Chicago Syphilis Control Program can also be viewed as a story, full of human characters who inform its goals, strategies, dilemmas, and triumphs. Like history and rhetoric, though, stories depend on words, and words are often capricious. The meaning of any word can reflect its moment in time, the intent of its writer or speaker, and (deliberately or unconsciously) the personality of its writer or speaker. Furthermore, anyone who attempts to retell or interpret other people's words or stories adds another layer of words and intentions to the telling. One way to guard against this inherent relativity is to collect as many different words or stories about an event as possible. To that end, this book presents government documents, newspaper coverage, and the day-by-day critique of one employee of the Chicago Syphilis Control Program. Thus, it offers three different perspectives on the same story: an "official" perspective; a popularized, attention-seeking perspective; and an insider's personal perspective.

Advocates of both the medical and moral interpretations of syphilis have always coexisted, sometimes collaboratively and sometimes contentiously. On the federal level, for example, the policies and practices of the U.S. government in its various efforts against venereal disease are inseparable from federal policies and practices around female prostitution. On the local level, Chicago's crusades against venereal disease before 1937 also often took this approach. And they always did so with the sensationalism and swashbuckling that characterized much of the city's growth, including dramatic exposés of such rackets as medical quackery and prostitution. At the same time, though, a significant number of Chicago's police, politicians, and public health officials were willing to consider cooperative relations with Chicago's underworld, an openness that would prove contrary to federal policies in 1937.

One of the city's major newspapers, the *Chicago Tribune*, had made a name for itself as an outspoken supporter of the city's earlier campaigns against syphilis. The newspaper prided itself on being a leader in breaking the taboo against the disease by printing the word *syphilis* in its pages, a forthrightness congruent with its philosophy of serving the public good. In keeping with its proud tradition, the *Tribune* willingly became the only

Chicago newspaper to carry daily coverage of the aims and activities of the Chicago Syphilis Control Program. Ironically, however, by placing the fight against syphilis within President Roosevelt's New Deal, the Health Service put the *Tribune* in an awkward position. An unremitting opponent of Roosevelt and his programs, the *Tribune* found itself increasingly caught in editorial tensions between its local (relative) progressiveness and its national conservatism.

Beyond national, city, and business institutions, the story of the Chicago Syphilis Control Program is a story of individuals. It boasts a cast of thousands, many of whom led illustrious, colorful, and notorious careers. On the national level, there were Drs. Vonderlehr and Wenger, whose background with the Tuskegee program must be reckoned with in analyzing Chicago's work in syphilis control. There was also Surgeon General Thomas Parran himself, who proclaimed syphilis a "shadow on the land" and began the national war against that disease. In Chicago, Herman N. Bundesen and Louis Schmidt were frequently in the midst of political and medical controversy. Bundesen, moreover, had just completed an unsuccessful bid for governor as the Roosevelt-backed candidate of Chicago's Democratic machine.

Among all of these prominent, colorful people, one of the most colorful, whose prominence was unique, was Ben Reitman. A high school dropout who rode the rails in the late 1800s as a hobo, Reitman subsequently entered medical school. He brought to his medical practice a familiarity with poverty, lawlessness, social defiance, and hopelessness—and the many physical ills that preyed on those conditions. As the lover of the anarchist leader "Red Emma" Goldman, Dr. Reitman learned an ideologic vocabulary to dignify his own often idiosyncratic beliefs. When the stormy relationship ended, he returned to Chicago with a criminal record for openly advocating and teaching methods of birth control, which was illegal at that time. During his medical career, Reitman worked with nearly every major local health official who would eventually become active in the Chicago Syphilis Control Program. His friendship with many of the program's leaders, his medical and personal past on the fringes of so-called respectable society, and his irrepressible enthusiasms and antagonisms offer a special perspective on the events and policies of the program.

Finally, books not only have historical or rhetorical contexts, but they also generally have—or acquire in the process of research and writing—personal ones. My interest in the Chicago Syphilis Control Program grew from my earlier work about the life of Ben Reitman, but my decision to write about the program came from similarities I saw between the issues surrounding the antisyphilis campaign of the late 1930s and the issues surrounding the treatment of AIDS and the HIV epidemic fifty years later.[24] Syphilis, obviously, is not AIDS. Not only are the two disease processes different, but also syphilis in 1937 did not report the mortality rate that characterizes AIDS today. Furthermore, cumbersome and uncertain as it was, treatment for syphilis was available in 1937.

Syphilis as a killer, however, was a major image in the antisyphilis campaign of the 1930s, as was the image of syphilis as inherent in illicit sexual promiscuity—and of promiscuity as inherently evil. Moreover, with both syphilis then and AIDS/HIV now, moral "solutions" have frequently been offered as the way to rescue the nation from insidious communicable disease. Reading the present into the past is fraught with logical—and scholarly—dangers, but considering the present in the light of the past may give insight into current debates. Such a consideration and speculation will become the subject of the Epilogue, because issues of fear, moral judgment, and responsible action—issues that underlie, pervade, and emerge from all events surrounding the Chicago Syphilis Control Program—form the basis for a critique of the rhetoric and images in today's campaigns against AIDS.

CHAPTER ONE

Taking the Pledge

In June 1937, six months after the national Conference on Venereal Disease Control Work, the Chicago Syphilis Control Program launched its attack against syphilis on three fronts: popular, legislative, and legal. On the popular front, the program began a highly visible campaign to garner the support and—even more important—the participation of the citizens of Chicago. Mayor Edward J. Kelly's kickoff in January, with speakers from many walks of Chicago's public life, exemplified that strategy and set the tone of civic zeal that would be a hallmark of the campaign. "This important meeting will go down in history as one that brought great benefit not only to this city, but to the whole United States," Kelly proclaimed, followed by Health Commissioner Herman N. Bundesen's characteristically dramatic words: "A campaign of education, of pitiless publicity, will break down the conspiracy of silence [Surgeon General Thomas Parran's frequently quoted term] and the public will respond as soon as the facts are known."[1]

Toward this end, the Chicago Board of Health mailed a survey to a million Chicago households, asking people if they would be willing to have themselves and their entire family tested for syphilis. Mailings from Chicago's Board of Health were not new; past campaigns about drinking water and baby care had often included sending educational material to many Chicagoans. In 1925, in fact, Commissioner Bundesen had sent a letter of "warning" about venereal disease to five hundred thousand Chicago households.[2] The syphilis survey, however, was the most ambitious of such undertakings to date.

On stationery bearing the letterhead of the Treasury Department,

United States Public Health Service, and embossed with the seal of the Office of the Chicago Board of Health, Bundesen addressed his constituency as both president of the Chicago Board of Health and special consultant to the Public Health Service:

> A nation-wide campaign is being conducted to wipe out syphilis. Federal, State, and Local agencies are cooperating in the Chicago program.
>
> More than one-half million men, women, and children in the United States become infected with syphilis every year, many of them innocently. Thousands of additional persons have the disease but do not know it.
>
> To keep some one near and dear to you from possibly getting this disease innocently, its source must be known. The support of every citizen in Chicago is necessary if syphilis is to be eradicated.
>
> The attached questionnaire is designed to register your protest against this disease. Will you please fill out the questionnaire, enclose it in the self-addressed, POSTAGE-FREE envelope, and MAIL it today?[3]

Bundesen's letter carefully emphasized the innocence of many of the people who contract, or might contract, syphilis. Moreover, he couched his request for voluntary testing in terms of protecting others, particularly "some one near and dear to you." It is important to notice the deliberate avoidance, in this initial contact with the public, of any language that would directly imply guilt or blame.

Bundesen's questionnaire was equally brief and to the point:

HELP TO PLACE SYPHILIS UNDER CONTROL

1. In strict confidence and at no expense to you, would you like to be given, by your own physician, a blood test for syphilis?
   Yes ______ No ______ (Please check one)
2. Including yourself, indicate on one of the two lines below, the age of each member of your family who would like to have a blood test:

   (A) The age of each male who would like to have the blood test:

   ______ ______ ______ ______ ______ ______ ______

   (B) The age of each female who would like to have the blood test:

   ______ ______ ______ ______ ______ ______ ______

PLEASE DO NOT SIGN YOUR NAME[4]

Urgency, voluntarism, and confidentiality—three major themes of the Chicago Syphilis Control Program—were sounded succinctly but clearly. To create a sense of urgency without implying impending doom, to ask for responsibility without implying blame, and to promise secrecy without implying shame is an ambitious undertaking. It was particularly ambitious, though, because past efforts at syphilis control often had incited doom, blame, and shame. The new syphilis control campaign brought with itself a new public relations strategy.

But old ways die hard. Indeed, Mayor Kelly's own words in January painted exactly the picture Bundesen so assiduously tried to avoid: "The danger of venereal disease lurks in every corner of this great city," he warned. "The time is here when every child of reasonable age should know what is in store for them unless they live according to the rules of Nature, and are careful about everything they do with reference to hygienic conditions."[5] The contrast between the innocent and guilty in Bundesen's and Kelly's language in the two announcements illustrates two approaches that had jostled for prominence in the public rhetoric about syphilis in the United States for more than seventy-five years.

Syphilis Control and the Federal Government

Bundesen's earlier letter about syphilis, in 1925, was unique for its time; by 1937, syphilis still was not considered a fit subject for polite conversation. When public discussions of syphilis were first heard in the late 1800s, they most often took the form of moral debate or proselytizing. The government did not take a role in that debate for many years. As Allan M. Brandt points out in his important history of venereal disease, *No Magic Bullet*, venereal disease was excoriated most often for its threat to the family.[6] Physicians, soon followed by religious, reform, and philanthropic groups, were among the first to write and speak about what they considered to be the personal and public dangers of venereal disease. They described in damning detail the spread of infection to such "innocent victims" as wives and children through the promiscuity of prostitutes and the men who visited them both before and after marriage.

Some of these groups favored public education. One, the American Society for Sanitary and Moral Prophylaxis, censured the government for confiscating and ruling obscene a pamphlet by Margaret Sanger, "What Every Girl Should Know," for its frank discussion of birth control, syphilis, and gonorrhea.[7] The Comstock Law, which forbade sending obscene materials through the mails, was invoked against Sanger's pamphlet. The law was representative of early U.S. legislation that directly or indirectly addressed venereal disease. On the whole, these laws were punitive and exclusionary. The 1891 Immigration Act forbade entrance into the United States of "persons suffering from a loathsome or dangerous contagious disease," a restriction the Public Health Service interpreted to apply to immigrants with venereal infections.

The American Society for Sanitary and Moral Prophylaxis opposed restriction and punishment as the primary solutions to venereal disease, preferring an educational approach. Although education was a step forward from exclusionary approaches, the society usually chose its words deliberately to instill fear of and revulsion for the diseases and the sexual profligacy with which they were usually linked. Educators and reformers hoped that this fear would lead men and women to shun extramarital sexual activity and embrace the chaste, monogamous marriage of virginal men and women. The term *moral prophylaxis* referred to sexual continence outside of marriage, making explicit its advocates' condemnation of extramarital sex.[8]

Setting moral standards for sexual behavior was central to the goals of another group of reformers at the turn of the century—crusaders against prostitution. These groups, which included physicians, members of the clergy, moral reformers, and philanthropists, eventually united with anti-venereal disease campaigners to become the American Social Hygiene Association. Opponents of prostitution used the increasingly authoritative voice of medicine, with its dire reports of venereal disease, to support and justify their efforts.[9] Although many of the reform groups tried to treat prostitutes sympathetically, believing that compassion and a good example were all some women needed, municipal authorities usually took a less beneficent approach. For the medical and moral safety of the American family, commissions on vice formed in large cities around the country and urged passage of laws requiring the arrest, examination, treatment, and, if necessary, quarantine of prostitutes. Other reformers investigated the economic, social, and moral milieux of prostitution with the sympathetic hope of being able to help prostitutes change their way of life. Still others felt that the best deterrent to prostitution was men's fear of contracting a venereal infection from prostitutes.

The latter group sought to create fear of venereal infection as a deterrent to sexual activity outside marriage. Treating a prostitute's venereal disease, they argued, would not deter men from visiting prostitutes. This argument continued to be heard, eliciting criticism from many feminists, even after the first reasonably effective treatment for syphilis was developed. Supporters of such measures insisted that men who caught a venereal disease from prostitutes had only themselves to blame. Although male customers as well as female prostitutes (and prostitutes tended to be seen and presented exclusively as women at that time) shared the blame for spreading infection, it was prostitutes who came to embody, both physiologically and metaphorically, the moral corruption assigned to extramarital sex.[10]

Prostitution in European countries at this time was frequently regulated by a country's government—for example, restricting prostitution to certain areas of a city or requiring prostitutes to meet certain health standards. Similar suggestions or efforts in U.S. cities were strongly opposed by reform groups, an opposition national leaders were forced to face as the United States entered World War I. Before that time, a skirmish between U.S. troops and Pancho Villa in New Mexico had vividly demonstrated to war officials that the proximity of alcohol and prostitutes to fighting soldiers resulted in high rates of venereal disease. Army physicians responded to that incident by conducting frequent medical examinations and providing chemical treatment for men after sexual encounters in the form of an ointment that, if applied early enough, would prevent the spirochete from doing its work. The use of *chemical* prophylaxis distressed representatives of the American Social Hygiene Association and the YMCA, staunch supporters of *moral* prophylaxis, who were on hand to observe the actions of the troops. In their eyes, providing chemical prophylactics flew in the face of the moral reform that lay at the heart of their programs.[11]

This ongoing debate over chemical prophylaxis was presented to Secretary of War Newton D. Baker, who responded by making a plea to the army to bring American boys home from the war in Europe not only clean but also morally untainted. To this end, activities such as baseball games, social clubs, and musical entertainment were provided for soldiers on bases in the United States as well as overseas. In addition, they received lectures about the shame, disgrace, and disaster of contracting a venereal disease or—even more reprehensible—the calamity that would result from bringing syphilis or gonorrhea home to loved ones.

In 1917, the General Medical Board of the Council of National Defense passed a resolution that held sexual continence to be healthful both for the prevention of venereal disease and as an expression of manhood, a statement expressly designed to counter arguments that sexual activity was a necessary component of virility.[12] At the same time, however, the

Armed Services was giving its men frank lectures on venereal disease, and, although it continued to urge soldiers to practice sexual continence, it also regularly examined them for venereal infections. In 1918, General Order No. 32 named the failure to use chemical prophylaxis after sexual intercourse a neglect of duty. Although there was strong opposition to these practices from outside the Army Medical Department, the "battle against [chemical] prophylaxis," concludes Allan Brandt, "was, for the duration of the war, a lost cause."[13]

When the war ended, the special federal board created to protect soldiers from venereal disease was dismantled. The American Social Hygiene Association, in turn, shifted its campaign for moral prophylaxis back to the civilian population. State boards of health were unwilling to make chemical prophylaxis a part of their treatment programs, and many states became uncomfortable with *any* public discussions of venereal disease. Several states banned, on grounds of obscenity, the showing of the War Department's—or any organization's—antivenereal film. In 1924 a poll of state health officials by the Public Health Service's Venereal Disease Division revealed that most states were unwilling to advocate chemical prophylaxis because of strong public opposition.[14]

Such was the situation when Dr. Thomas Parran became chief of the Venereal Disease Division of the Public Health Service in 1926. Although he agreed with the opponents of chemical prophylaxis, he strongly disagreed with the policymakers who would keep silent about venereal disease. By the time Parran became surgeon general and head of the Public Health Service, however, syphilis was once more seen as a sizable health problem, and Parran found himself in a position to speak up.[15] In an article entitled "The Next Great Plague to Go," printed first in the *Survey Graphic* and reprinted in *Reader's Digest* in 1936, and then in his book *Shadow on the Land: Syphilis* a year later, Parran outlined his eradication plan: locate every case through mass testing and tracing the sexual partners of infected people, treat all cases, and educate both physicians and the public.[16] The same goals prefaced the first annual report of the Chicago Syphilis Control Program and were referred to in all its official documents.

Parran justified the magnitude of such an effort on the utilitarian grounds of public safety, economic savings for the nation, scientific opportunity and responsibility, and official duty. "As a federal health officer," he wrote in the preface to his book, "it is my task to save lives by applying medical knowledge against the plagues which afflict the most of us, and the most dangerously; to save taxes by reducing the number of physically and mentally unfit who crowd our public institutions and swell the proportion of the unemployable in the population; to promote research for better methods of doing things; and to report the truth as I find it to citizen and official alike."[17] By casting syphilis as a public health problem rather than a moral one, a matter of sound business economics rather than family righteousness, and an opportunity for scientific advancement rather than an outrage against social purity, Parran created a new realm for governmental action. By using language that resonated with current public views about the nature of civic responsibility for both elected officials and public citizens, he linked individual health and behavior to the community good.

In much of his writing, Parran presented syphilis the disease (as opposed to people infected with syphilis) as the enemy. Such an approach is consonant with the posture of a public health officer responsible for checking the spread of infectious disease as well as directing approbation to community rather than individual responsibility. In his first annual report after the inauguration of his campaign against syphilis, Parran's words echoed Herman Bundesen's belief in the liberating potential of scientific fact: "The hampering, ostrich-like attitude toward these diseases is gradually being overcome. When they are brought out into the open, freed from the medieval concept of condign punishment for moral transgressions, and dealt with as are any other highly communicable diseases, the way is open to eradicate them just as we have stamped out other dangerous infections." Parran and others were attempting to remove the personal stigma so long attached to syphilis by declaring it a community threat in the same "ranks [as] cancer, tuberculosis, and pneumonia."[18]

Such a depersonalization, however, is not necessarily objectification, nor does it necessarily follow that this official, rhetorical reframing of syphilis and its consequences either reflected or immediately changed generally held attitudes or beliefs. This is not to say that language carries no influence over thought or perception, but in this instance Parran and his followers were asking most people to think about syphilis in an entirely new way, to adopt a new metaphor almost overnight. Parran himself was not entirely free of old ideas; he often presented his argument in a way that lay blame for the spread of syphilis on unresponsive individuals and communities. Such an implication echoes the earlier stand of the reformers who discouraged chemical prophylaxis on the grounds that profligate men get what they deserve. Now, entire communities could be punished for social irresponsibility. As Parran put it, "Syphilis is a large though undefined factor in the problems of the home, the community, the state, and the nation. Because of it marriages are prevented, families broken up, children born dead or handicapped. It lessens the efficiency of the young and vigorous. It shortens life. It adds to the public burden of the physically and mentally unfit." He described a familiar picture of family dissolution and physical and mental devastation. "What the individual must do about it and what the community should do are inextricably interlinked," he insisted. "Both forms of action are urgent. Both have been neglected."[19] The personal and the public *were* "inextricably interlinked." Therein lay many of the problems that the Chicago Syphilis Control Program would encounter.

"My Whole Family Will Gladly Submit"

The Chicago Syphilis Control Program was a culmination of events that had made syphilis a public concern and responsibility. Chicago, however, was not the only city—nor was it the first—to enter into such a partnership with the Public Health Service. Baltimore and New York both had active programs, Baltimore's since 1931. In Baltimore, Surgeon General Parran announced that Maryland would become a leader in the fight against "social diseases."[20] By June of 1937, New York already had a reporting system in place, and patients were receiving free treatment through the Social Security Administration.[21] During this period, the *Chicago Tribune* recorded the rising sense of the urgency of problems created by syphilis. In "Asks for Added Health Powers in Syphilis War," the *Tribune* printed a story from Atlantic City that described the dearth of facilities and programs nationwide for detection and treatment of syphilis.[22] A second story, "Urges Inclusion of Gonorrhea in Syphilis Drive," followed three days later. One of the few stories about gonorrhea ever to run in the *Tribune* during the time of the Chicago Syphilis Control Program, it reported on an unsuccessful trial of sulfanilamide treatments for gonorrhea in Baltimore. It concluded by urging that a system be devised for testing, treating, and teaching about gonorrhea similar to the one just devised for syphilis.

Soon, however, the paper turned its attention closer to home, and on June 14 the *Tribune* announced the upcoming syphilis survey. "All Chicagoans Urged to Take Syphilis Tests" called out a front-page headline.[23] The theme of mutual responsibility began to sound ever more clearly. On Sunday, June 20, Presbyterian minister John Evans was quoted, "The prevalence of such diseases [as syphilis] is in part due, if not to a conspiracy of silence at least to a timidity, which in the interest of truth, health, and the wholesome development of our young people must be overcome."[24]

The Reverend Evans's words contain one of the major tensions that would persist throughout much of the rhetoric surrounding the Chicago Syphilis Control Program. The service of health enlisted alongside the service of truth immediately invites semantic and ideological debate. Add to those goals the wholesome development of the nation's youth, and a confrontation of values is inevitable. Such a paradox emphasizes the challenge of making personal issues—that is, truth, wholesomeness, and even health—public. It is, moreover, the same conflict inherent in Parran's insistence that syphilis is both a public and private responsibility.

The Chicago Syphilis Control Program, however, promised to bring in a new era in the treatment of syphilis. That Chicago had been chosen as the flagship city in the U.S. Public Health Service's antisyphilis campaign, moreover, was made emphatically clear. "Health Service Hails Chicago's Syphilis Battle," proclaimed an article in the *Tribune* in late July 1937. "First Realistic Attack" Surgeon General Parran said in praise of the program.[25] Chicago's approach was to be a positive, upbeat one. It would make personal and public responsibility for syphilis a source of pride, celebratory rather than punitive. Mailing the syphilis survey was treated as an invitation to a public gala—or, perhaps more accurately, a masked ball. "Free and confidential" was a rallying cry for the event. A copy of the survey was printed in one issue of the *Tribune*, and a series of pictures on its photo page outlined the testing procedure. Much ado was made of the million surveys. On the front page of the same issue, the *Tribune* described the schedule for mailing the questionnaire: 250,000 the first day, then fifty thousand surveys from Bundesen's office a day until the full million were sent. In addition to the board's mailing, Dr. Rachelle Yarros, a leading Chicago physician, reformer, and academician, arranged through the Urban League to have special postcards from Commissioner Bundesen sent to 350,000 mothers in the city.[26]

At the same time, another article announced that members of Chicago's Lions clubs had adopted a resolution to help spread information about syphilis and its control.[27] It was the first of many such stories that ran throughout the early months of the Chicago Syphilis Control Program, reporting the various professional, fraternal, or civic groups that had announced their support of the program and the intent of members to be tested. At many gatherings, free tests were available at the meeting's close.

Among the groups that so pledged or were tested were a Chicago cooks' union, a group of county club women, the Polish-American University Club, and American Legion posts.[28]

Such stories continued to appear in the *Tribune* for more than a year. In late March of 1938, 65 percent of the paper's staff was tested. Of these 1,755 men and women, only 50, the paper reported, chose to be tested by their private physicians. The others were tested at the *Tribune* offices by two physicians who worked for the Chicago Syphilis Control Program. The description of the event was carefully worded to present the activity as a fraternal gesture of goodwill, an exhibition that carried no shame, as the Chicago Syphilis Control Program itself insisted. The staff, the story read, took the test under "no pressure . . . [and] to their personal advantage."[29]

Endorsement by health professionals' societies was sought and gained early on. Concern about backing from the politically powerful and conservative Chicago Medical Society had been voiced early on at the national level, but in July 1937 the *Tribune* reported the society's support, noting a similar endorsement by the Chicago Dental Society a few days later.[30] Stories of growing support for the testing program mingled with almost weekly tallies of the numbers of questionnaires that had been sent and returned. "Younger Women Willing to Take Syphilis Tests; Poll Results Encouraging, Says Dr. Bundesen," read the headline of a story on August 27, reporting that women between the ages of fifteen and thirty-four had so far been the group most willing to be tested. By August 6, responses were reported to favor free testing 70 to 1.[31] By the end of August, a 10 percent return on the million questionnaires showed that 95 percent of those responding favored the test. The mandate was so clear to the Chicago Board of Health that its members decided to begin testing on September 1, a month earlier than planned.[32]

Even in an anonymous pool of one million confidential questionnaires, the *Tribune* was able to discover a human interest angle. On August 21, the newspaper reported that the recipient of the millionth questionnaire, Norman Plambeck, a butcher with a wife and two children who resided at 4226 N. Meade Avenue, had returned his questionnaire, signed with the following words, "My whole family will gladly submit to the tests. Every citizen should cooperate in such a worthy campaign, and particularly those who are in the business of distributing foodstuffs to the public."[33] Despite Plambeck's public-spiritedness, health officials knew by this time that it was extremely unlikely that syphilis could be conveyed through food handling, yet they still broached the possibility cautiously. Correspondence from R. A. Vonderlehr in the surgeon general's office in 1938 recommended that persons testing positive for syphilis should not be denied employment, but people with "open sores . . . should be temporarily excluded from employment until such lesions have healed." He went on, however, to observe that anyone testing negative at one time may still "acquire syphilis shortly thereafter," thus making the test primarily useful only "as a method of case finding" but of little "practical value from a standpoint of protection to the public."[34] None of the debate appeared in the *Tribune* article, however, and the story ends, simplistically, with a warning to the public of the dangers of anyone infected with syphilis handling food.

Regional Consultant O. C. Wenger reported the results of the survey in his first annual report of the Chicago Syphilis Control Program. Of the million ballots mailed, 99,000 were returned. "Yes" ballots numbered 93,931, as opposed to 5,202 "no" votes. For a mass mailing, this response was considered good. Totaling the number of people per household on the "yes" ballots, 261,425 people indicated their willingness to be tested for syphilis. Wenger concluded with unqualified optimism:

> This large response is an effective rebut[t]al to the often expressed idea that the public is not interested in the syphilis problem and would show no response to a request of this kind. It was as a result of this poll that the extra $50,000 was obtained for laboratory procedures. We now have the basis for a city-wide survey, by which it is hoped we may obtain, for the first time, figures regarding the real incidence of syphilis in the American population as a whole.
>
> The attitude of the public, as shown by this poll, is one of the most encouraging factors in the entire program. The large increase in reports from private laboratories would also indicate that thousands of persons who did not return the ballot went to their own physician on receipt of the questionnaire and had a blood test.[35]

Bundesen's survey thus became the means (or the justification) for even more federal money to be directed to the Chicago project and encouraged federal officials that public response would be sufficient to generate information of use to the nation. Moreover, officials were impressed by the apparent enthusiasm of people who would be prompted by the survey alone to apply for a blood test. Chicago seemed to be a charmed city for syphilis control.

Syphilis Control and Medicine in Chicago

By summer 1937, then, Chicago had ostensibly gained support for its Syphilis Control Program from the medical, moral, and civic communities. Such apparent unanimity was remarkable, even given Chicago's long history of "advanced" views about syphilis. A brief look, however, at the history of public health in Chicago reveals the myriad political, social, and personal issues that percolated beneath this deceptively harmonious surface.

Thomas Neville Bonner, noted historian of public health in Chicago, locates the origins of the city's public health work in efforts similar to those of many U.S. cities: in responses to epidemics or other emergencies.[36] In Chicago, outbreaks of smallpox or cholera in the 1840s and 1860s and the Great Fire in 1871 spurred the first public health efforts. The board of health that was established during the first crisis was abolished during financial depression in the mid-1850s, then reestablished in 1867. At that time, its major responsibility was the sanitary inspection of streets and buildings, another common phase in the evolution of city health departments. When Cook County Hospital opened in 1864, responsibilities for health care were divided: The county would care for the sick poor, and the city would care for people with contagious diseases, take preventive measures against epidemics, and record the city's vital statistics. All of these responsibilities of the Board of Health would loom large in the Chicago Syphilis Control Program years later.

The development of germ theory and resulting immunization techniques further heightened the effectiveness and status of public health beyond sanitation responsibilities. Bacteriology made its appearance in the Chicago Health Department in 1893 with milk inspection. The success of that program led to a municipal laboratory headed by Dr. Adolph Gehrmann, who quickly expanded his work to include testing for diphtheria, typhoid, and rabies. The department soon began to conduct public campaigns against typhoid, summer health problems of babies, and infant mortality. Education was frequently a part of the program, which often involved mailing public service material, such as Bundesen's letter about syphilis in 1925, to residents.

Around the turn of the century, the Health Department began what was to become a continuous struggle to convince physicians to report many communicable diseases. Reporting had been required by law in Chicago since the 1870s, but private physicians felt the city should reiburse them for filing such reports, citing as precedent the 25 cents they received every time they filled out a birth report. Such noncompliance, although not unique to Chicago physicians, became a recurring complaint from the Board of Health and a particular frustration during the syphilis control program of the 1930s. This and other disagreements between public health officials and private physicians have peppered the history of the Chicago Board of Health. On one occasion, Dr. William A. Evans, one of Chicago's more notable health commissioners (1907–11), aroused the wrath of the Chicago Medical Society for enacting the country's first ordinance for mandatory pasteurization of milk. The society argued, first, that such laws would undermine the sanitary measures currently used in milk production, and, second, that buyers of milk should be able to choose "live" rather than "cooked" milk if they so desired.

Among other firsts that Dr. Evans achieved during his brief tenure as health commissioner was the opening of the city's first venereal disease clinic in 1910. The clinic soon closed, for reasons not recounted by Bonner, but in 1911 under Evans's direction the Health Department added Wassermann tests to detect syphilis to its regular services. In that same year the city abolished its segregated "vice district," in what Bonner calls Chicago's "first serious attempt at the control of venereal disease." The city's first effort was thus criminal rather than medical, an approach in keeping with the actions of many cities throughout the country at the time.

New medical efforts of control came after World War I, when the extent of venereal disease in Chicago and a successful means of treating it both became widely known. The city immediately launched two programs, one public and one private. The public treatment program operated in the city's Municipal Hygiene Clinic, begun by Health Commissioner Dr. John Dill Robertson. During Robertson's tenure as commissioner (1915–22), Chicago became, in 1917, the first city in the United States to pass ordinances that required reporting and treating all known cases of syphilis and gonorrhea. In 1919, another new ordinance permitted judges to order tests "for persons suspected of having venereal disease and authorizing health authorities to segregate them for treatment."[37] Although it provided syphilis testing and treatment for any Chicagoan who sought it, most of the Health Department's energies were directed toward the inmates of Chicago's city and county jails, a policy that would unintentionally reinforce associations of criminal behavior with venereal disease.

The second, private, treatment program in Chicago was the Public Health Institute (PHI), a nonprofit clinic whose staff diagnosed and treated only venereal diseases, charging fees well below the usual cost.[38] At this time, there were 154 such clinics in the United States. The PHI was founded in 1919 by two men who had served in the U.S. Army during World War I. Joseph G. Berkowitz had been a physician in the Army Medical Corps and Myron E. Adams had served as a morale officer. Both had worked in the army's venereal disease programs, an experience that convinced them that civilians could also be taught the seriousness and treatability of venereal disease. The PHI, however, was a treatment center; it did not teach its patients about chemical prophylaxis.[39] Thus, Berkowitz and Adams aligned themselves with the groups that advocated moral rather than medical measures of prevention.

Within ten years the PHI was the largest clinic of its kind in the world, and, much as Chicagoans enjoy superlatives, some citizens had trouble with this one. The Public Health Institute operated on the belief that a private business could provide low-cost treatment to all people who needed it, a proposition that aroused the capitalistic ire of the Chicago Medical Society. Although the society couched its criticisms in other arguments, Conrad Seipp, a historian, contends that the organization feared that individual practitioners would lose patients to the much less expensive physicians and services of the PHI—although in fact relatively few Chicago physicians treated either venereal diseases or the poorer patients who comprised the bulk of the institute's business.

The full anger of the Chicago Medical Society erupted in 1929, when the Public Health Institute contracted with the Illinois Social Hygiene League (the state chapter of the American Social Hygiene League). The league, which had been the primary organization to educate the public about venereal disease before the PHI opened its doors, offered to pay the PHI to provide free treatment of venereal disease to indigent patients. Some of the city's most notable social and medical reformers belonged to the league. Of special note were Jane Addams, founder of the settlement system at Hull-House; Rachelle Yarros, the first woman professor of obstetrics and gynecology in a coed medical school (now the University of Illinois) and founder of Chicago's nine birth control clinics;[40] Herman Bundesen, former (and soon to be reinstated) commissioner of the Chicago Board of Health (1922–27 and 1930–68); and secretary and president of the league Lewis Schmidt, appointed in 1903 to a committee of the American Medical Association on prophylaxis of venereal disease, the former director of venereal disease control for the Chicago Health Department, and a member of Chicago's Vice Commission of 1911.

It was upon Schmidt that the wrath of the Chicago Medical Society fell.

He was charged with violating society rules and those of the American Medical Association regarding advertising, and he was expelled from membership in the local society. The fight that followed drew not only local and state but also national attention. It was further fueled when Herman Bundesen, then serving as city coroner, resigned his membership in the Chicago Medical Society in protest. The *Tribune* and other local newspapers ran numerous stories and editorials attacking the society for trying to keep low-cost medical care from the public despite the society's insistence that theirs was an ethical and not an economic grievance against Schmidt. Yet Schmidt's career was not damaged irreparably by the expulsion. By 1937, he was again working closely with Bundesen on the Chicago Board of Health, in charge of the Division of Venereal Disease—and causing anxiety to Vonderlehr and his Washington colleagues.

Chicago's first *city*wide drive against venereal disease began a few years after the founding of the Public Health Institute. One of the first major projects that Bundesen undertook upon being named commissioner of health in 1922 was the eradication of venereal disease from Chicago. Like Berkowitz and Adams of the PHI, he too had learned the lessons of war, but Bundesen, following more the philosophy of public health than of moral reform, took away a different lesson. He chose utilitarian over utopian approaches. "The training received by the men in the army has borne fruit," he argued. "'It is better to be safe than sorry,' is appreciated by many."[41]

Bundesen's focus in his campaign, in keeping with most city and military campaigns to date, was on prostitution. He raided houses of prostitution with a vigor that hadn't been exercised in Chicago in years, using to advantage the existing laws that allowed mandatory testing of people arrested and the segregation of infected prisoners. By 1926, "All prostitutes suffering with an active, contagious venereal disease were promptly sent to Lawndale Hospital," the city's infectious disease hospital. When such measures failed, houses where untreated, infected prostitutes worked were placed under quarantine, and "the prospective customer [would] be greeted by a large glaring placard, 'Venereal Disease Here—Keep Out.'"[42]

Bundesen also adopted a new approach—new, at least, for many U.S. cities. Part of his pledge to "drag the snake out of the bushes" entailed meeting with "vice lords" of the city's underworld. The *Chicago Tribune* outlined Bundesen's goal to include a prophylactic kit "in the price paid by patrons of brothels" and to have the houses' "women inmates instruct men patrons in the use of the sets." The kits contained condoms, creams, and tablets that were to be dissolved in water for douching or washing.[43] In 1926, the Division of Social Hygiene reported, "In Chicago, Commissioner Bundesen's policy that 'if you won't be moral you must be clean' has been instrumental in psychologizing the 'keepers' of 'houses' and the inmates into taking extra precautions to avoid venereal disease."[44]

Bundesen's program, however, backfired, although whether because of its zealousness, harshness, emphasis on prevention, or mere visibility is unclear. Bonner reports that Bundesen's actions brought an immediate "storm of public disapproval"; nearly every local newspaper, including the *Tribune*, attacked Bundesen, as did nearly every medical journal in the city. He was forced to retreat from many of his projects, and Bonner concludes that "realism, Bundesen learned, was not the order of the day."[45]

Certainly, for whatever other reasons Chicagoans objected to Bundesen's program, his realistic approach to prostitution and venereal disease was a new one for most public health departments to pursue, at least so openly. Whereas in earlier decades raids on houses of prostitution were ostensibly intended to end the profession itself, Bundesen's actions, as a means of finding and treating syphilis, were an open acknowledgment that prostitution was a part of daily life. His main concern lay in keeping prostitutes and their customers as healthy as possible—for the public good.

A pertinent footnote to this early program comes from another episode in Bundesen's career, this one occurring during his three-year absence from the leadership of the Board of Health. In 1927, he was dismissed as commissioner by Mayor "Big Bill" Thompson for refusing to include political literature within a mailing about baby care to all Chicago mothers.[46] With hardly a pause, Bundesen easily won election to another city office, that of city coroner—with Louis Schmidt, moreover, as his campaign manager. It is also clear that he considered this turn of events to be only a temporary hiatus. Bundesen's dedication to public health could not be sidetracked by even the demands of Chicago's gangland, as A. A. Dornfeld's tale from the St. Valentine's Day Massacre illustrates:

> Abe Lincoln Mahoney, another [City News] bureau reporter, was lurking about the scene of the crime, looking for a story, when he spotted the overcoats of the slain gangsters hanging on a long coat-rack on the wall of the garage. No one was paying any attention to him; although the place was swarming with policemen and reporters, most of them were concentrating on interviews with the coroner, Dr. Herman N. Bundesen, a popular pediatrician who had been trans-

ferred only recently from the Health Department. Mahoney spied a sheaf of papers in the pocket of one the overcoats. While everyone was distracted he quickly seized the papers and made off with them, hoping for a scoop. A few minutes later he slunk back into the garage and stuffed the papers into what he hoped was the right pocket. They were a sheaf of notes for one of Dr. Bundesen's articles on baby care.[47]

Why Bundesen fought for the integrity of his infant welfare program rather than his innovative venereal disease program could have had as much to do with the political times as it did with personal priorities. He was outspoken and shrewd, a man who often took extreme positions but who appears to have learned, early in his long career, either how to garner public support or how to choose his fights so as to emerge unscathed in the public eye.

"An Unlucky Day for Syphilis"

Certainly, in the early summer of 1937, Herman Bundesen was once again a hero, the leader of a public crusade that was deliberately focused on the hopes of American society. It was an artfully designed promotion. The Chicago Syphilis Control Program was making the world safe for innocent women and children. The call to be tested for syphilis took the tone of a temperance tract, and people who came forward to take the pledge were invited equally from among the sinners and the (as-yet) sin-free.

The presence of children in the camp further attested to the nobility of the venture. One of the most dramatic early events to draw support for citywide testing was a march of 1,500 members of the National Youth Association, who carried banners through Chicago's Loop on August 13, 1937. "Friday the 13th is an Unlucky Day for Syphilis" and "Chicago Will Stamp Out Syphilis" the banners proclaimed. The march ended at a city square, where the crowd heard speeches from Dr. George G. Taylor, *new* director of the Division of Social Hygiene of the Chicago Board of Health; Dr. Louis E. Schmidt (*now* secretary of the Board of Health); a statistician from the Public Health Service named Carsten; and the famous physician, crusader, and writer Paul de Kruif. De Kruif praised his young audience, saying, "It is you young fighters who have smoked one of mankind's most secret enemies out into the open."[48]

The day following the Youth March, an airplane flew across Chicago's sky, trailing a banner that urged people to send in their questionnaires: "Chicago, Fight Syphilis—Vote Today." The *Tribune* wrote of "the scourge that could not have been publicly mentioned a few years ago." It was, the paper and the Chicago Board of Health were quick to point out, a new—enlightened—day.[49]

CHAPTER TWO

The Clap Doctor of Chicago

On the day of the Youth March, at least one member of that enthusiastic crowd questioned the depth of the young audience's understanding. In one sense, that observer saw the children as pawns in a public-relations game, and he privately addressed them, "Yes, happy children, maybe it is just as well that you did not understand." He then went on to address a larger audience, the overseers of the Chicago Syphilis Control Program, saying, "And let us pray God that they'll never know what it means. But, all in all, it was an important event. The bankers and brokers and the insurance men looked out of their windows at the signs. Many of them knew what it meant. The stenographers and the shop girls from the sidewalk of Jackson and LaSalle streets read the signs, and they knew what they meant." The ends may have justified the spectacular means, he conceded; nevertheless, he wrote with some disgruntlement as he described his own presence at the rally: "An old man with long disheveled hair marched with the children down the street, and stood in the Square, listening to Bundesen and de Kruif. He was mumbling, and unhappy. [Carsten] had asked him to speak to the children at the Great Northern Theatre but his bosses told the chairman not to let the man speak." [1]

"Old man" Ben Reitman, then fifty-eight, was both a veteran and an organizer of similar marches. He was also a physician whose career spanned almost every aspect of syphilis treatment from the start of the twentieth century and who was acquainted, often closely, with nearly all of the city's medical and political leaders involved with the Chicago Syphilis Control Program. "Syphilis is enough for any man!" [2] Reitman said of his life, no small tribute to be given this disease, as Reitman's own life encompassed many causes and adventures as he rubbed shoulders with hoboes, anarchists, gunmen, prostitutes, scholars, and literati.

Reitman's lifelong identification with the underdog and his suspicion of all bureaucracies give his perspective on the Chicago Syphilis Control Program a surprisingly contemporary ring. It would be too easy, though, to proclaim him simply "a man ahead of his time." Reitman himself warned of the dangers of accepting one person's view unquestioningly. In the preface to his unpublished autobiography, he observed, "I am determined to be honest in this book, but that is a large undertaking. I find that I have mixed fiction and fact so much that it is no easy matter now to distinguish one from the other."[3] Reitman's views—as do those of every chronicler of the program—contain their own logic and illogic, unique to his dreams and desires, some of which may delight and some offend. Nevertheless, his perspective, contained in more than four hundred reports that he wrote while he worked for the program, offers a valuable window on its daily activities.

The Chi Kid

Benjamin Lewis Reitman was born in Minnesota in 1879. His mother, Ida Reitman, reared Ben and his elder brother, Lewis, alone after her husband left the family and went to New York. Upon their divorce in 1883, she moved with her sons to the South Side of Chicago, where Ben contributed to the struggling household by running errands for the many prostitutes who lived nearby. He was ten when he hopped his first freight train with school chums and discovered a passion for travel—paid or otherwise—that never entirely left him. He criss-crossed the country, learning the train yards and jails of most major American cities and talking to hoboes who tutored the young Chi Kid, as he came to be known, in the ways, talk, and life of the " 'bo."[4] When he was in Chicago between rides, Reitman worked at an assortment of jobs. Of these many stopgaps, he wrote, "I had one job in Chicago that lasted longer than the first thirty jobs I had put together. That was office boy for the Cook Remedy Company. This was a patent medicine concern that sold a treatment for syphilis."[5] The company's manager, C. L. Farnsworth, always rehired young Ben after his hoboing absences, and Farnsworth's wife urged him to attend night school at the YMCA, from which he eventually graduated.

In 1898, the YMCA sent Reitman to work as a laboratory assistant at Chicago's Polyclinic Hospital, a postgraduate institution offering courses primarily to Illinois physicians in private practice. It employed physicians whose main activities were teaching and research. Here, Reitman engaged the interest and support of several physicians, including the bacteriologist Maximilian Herzog, who was working on the etiology and pathology of syphilis and who, according to Reitman, performed one of the first Wassermann tests for syphilis on Reitman himself.[6] Also on the faculty of the Polyclinic were pathologist Leo Loeb, brother of the famous University of Chicago physiologist Jacques Loeb, and Adolph Gehrmann, who was instrumental in establishing the medical laboratory of the Chicago Health Department (chapter 1). When Loeb was offered a position in pathology at the College of Physicians and Surgeons (the "P&S," now the University of Illinois College of Medicine), he took Reitman with him as his assistant, stipulating that Reitman's salary, $100 a year, be used for tuition to that school. Reitman's application to medical school was accompanied by letters from Herzog; Fernand Henrotin, another famous Polyclinic surgeon for whom the Polyclinic Hospital was renamed after his death; and William A. Evans, a future commissioner of the Chicago Board of Health and also a professor of pathology at the P&S.

Reitman enrolled in the College of Physicians and Surgeons in 1900. He earned money to cover expenses by bringing in stray dogs to be used for teaching and research, cleaning the bones of dissected human cadavers to sell as skeletons, and later lecturing in Chicago dental and veterinary schools. "I had a very difficult time the first year," Reitman remembered. "I not only had to learn anatomy, physiology and chemistry, but I had to learn reading, writing, and spelling."[7] These hardships notwithstanding, Reitman found time during that year to marry a music student, May Schwartz, and travel with her to Europe, where he left his pregnant bride to continue her studies. He returned to Europe the next year to bring May and their infant daughter back to Illinois, but Reitman and Schwartz remained separated, divorcing soon after that. Around that same time, Reitman (along with several other classmates) was dismissed from the P&S on charges of cheating.[8] He immediately transferred to the American College of Medicine and Surgery (now Loyola University's Stritch School of Medicine), from which he graduated two years later. "Dressed in a Prince Albert coat and a bulging stiff shirt, surrounded by a proud mother, a happy brother and three lady friends," Reitman boasted, "I graduated from what is now the Loyola Medical College, in May, 1904. I passed the examination of the Illinois State Board of Health, received a license and immediately opened an office at 39th Street and Cottage Grove Avenue." Dr. John Dill Robertson, one of his professors at the college, loaned him $50 to pay for office rent and equipment.[9]

Who was this young Dr. Reitman? Tutored by hoboes, businessmen, and medical scientists, Reitman had, by his early twenties, already lived in several worlds but had probably not found a true home in any of them. He hid his efforts to learn basic reading and writing skills from his classmates in medical school, acutely aware of his differences from them. Whatever his academic handicaps, though, he captured the interest and won the sympathy of several influential Chicago physicians, whose support would continue, in some instances throughout his life.

Reitman established the first of his many medical offices among the kinds of people he had known most of his life. "From the day I opened my office I began to have patients—most of them were of the underworld type," he recalled.[10] His autobiography registers delight with this juxtaposition of the respectability of medicine and the taverns, houses of prostitution, and gambling dens that surrounded his office. Upon becoming a physician, Reitman did not completely abandon his footloose ways, however. During the first years of his medical practice, he frequently shut or even gave up his offices to hop a train or tramp steamer. He reported working, at various times, as physician to a boatload of European immigrants, victims of the San Francisco earthquake, workers on a stretch of Mexican railroad, and both the human and equine members of Buffalo Bill's circus on its European tour. Of the last of these jobs, Reitman commented, "I treated a large number of the boys of the circus for venereal disease. Nearly everybody working in the circus caught gonorrhea. The only ones that did not catch syphilis were those that already had it."[11] Reitman claimed to have visited the laboratories of Paul Ehrlich, Rudolf Virchow, and Elie Metchnikoff, men known for their contributions to the understanding and treatment of venereal disease, but there is no independent corroboration of these claims. After such forays, Reitman would return home, reopen an office in his old neighborhood, and begin his practice again.

Closer to home, in 1907 Reitman rode the rails to the hobo convention in St. Louis, an annual event of the International Brotherhood Welfare Association, an organization founded in the 1890s by John Eads How. How, the socialist son of a millionaire, was educated at Harvard and Oxford and believed that the hobo "had been driven to the road by industrial conditions over which he had little control." How's association was an educational movement, taking the form of "hobo colleges"—"the migratory worker's university"—in cities across the United States.[12] Upon his return from St. Louis, Reitman established a hobo college in Chicago. He spent the next four years, uninterrupted, in Chicago, his longest stretch at home since childhood. His medical offices housed the hobo college, where, Reitman remembered, hoboes eventually came "so thick and so fast that the tenants complained to the agent of the building, and I was asked to leave and take all of my hobos with me. The manager said, 'They are pestering this building like cockroaches.'"[13] Eventually, the Brotherhood Welfare Association had a meeting hall of its own. Reitman brought in speakers from neighboring colleges and universities to speak to and debate with the hoboes, organized special dinners for the homeless men and women, and conducted poetry readings and dramatic productions.

Reitman's growing involvement in the concerns of unemployed and often homeless men and women attracted public attention in January 1908, when he led a March of the Unemployed through the Loop. With painful memories of the Haymarket Riot twenty years before still generating antagonism and anxiety, the group's socialist leaders were on the verge of calling off the march minutes before it was to begin. Amid the growing confusion, Reitman stepped forward and announced, "There'll be a parade anyway. If the unemployed are there and want to parade, I'll lead them." As Reitman's biographer Roger Bruns describes it, "Swinging his walking stick at his side, a petition to the mayor and city council stuffed in his pocket, Ben strode over to the lakefront alone, just before two o'clock, to take command of the demonstrators. Dressed in a long black coat and slouch hat, smoking one cigarette after another, he hustled through the crowd giving instructions." They walked in pairs down Michigan Avenue, their numbers growing as they advanced. "As they neared Adams, the police charged through the ragged formation with billies flying. Ben defiantly pushed forward, his bearing erect, his walking stick still at his side, as the ranks behind him scattered in disarray, many bleeding from fresh wounds. . . .Emotionally and physically battered," Reitman was taken to the Harrison Street police station, where he was charged with disorderly conduct and released on bond.[14]

By 1908, then, Reitman had established a career that gave him a close acquaintance with, interest in, and commitment to people who were considered society's "outcasts," a term Reitman often used. Through both choice and circumstance, he often found himself in positions of leadership among them. The authority that accrued to his professional position may have carried some weight, but so, too, did his enthusiasm. His work with the hobo college showed his ability to organize people, and his day with the

March of the Unemployed his ability to rally them, two skills that would be honed even finer in the next decade of his life.

Red Ben

It was through Reitman's connection with the Brotherhood Welfare Association that he met Emma Goldman, the anarchist heroine-villain of the first decades of the twentieth century.[15] Emma Goldman was born in Russia in 1869 and emigrated to the United States in 1882. Inspired by the teaching of the anarchist Johann Most, she embarked on a life of political activism. Goldman and Alexander Berkman, for a time her lover and her lifelong friend, searched for ways to act upon their scorn for the corruption and injustice they found inherent in all bureaucratic, class-bound systems. Goldman came publicly into her own when Berkman was imprisoned after a failed attempt to assassinate industrial magnate Henry Frick, who ordered an attack on striking steel workers in 1892. She developed her own political philosophy of an anarchism that championed such causes as organized labor, free speech, women's suffrage, free love, the Russian revolution, birth control, and nonconscription. Her criticisms of government and industry infuriated and frightened many citizens, while her impassioned rhetoric and dynamic presence thrilled and inspired others—among them young Leon Czolgosz, who assassinated President William McKinley in 1901. Attempts to convict Goldman as well for the assassination failed, but by the time she met Reitman in 1907 Goldman was a nationally known figure of considerable controversy.

Reitman met Goldman when she came to speak in Chicago in 1907. Unable to rent a hall for her lecture, Goldman finally found a platform at the hall of the Brotherhood Welfare Association. Ben Reitman both extended the invitation and handled the arrangements. Storm and confusion surrounded the talk. It was disrupted by police, who had probably been informed of the event by Reitman himself, a clear sign to many of Goldman's supporters that Reitman was politically naive at least—and certainly not to be trusted.[16] This act and its tumultuous aftermath of anger, conflicting with a strong sexual attraction between Goldman and Reitman, set the tone of passion, repeated disappointment, and reconciliation that marked their ten-year love affair.

Soon after meeting Goldman, Reitman once more left his medical practice, this time moving to New York, where he was Goldman's business manager, lover, and comrade. His medical credentials gave special weight to his own lectures on sexology and birth control. Tutored by Goldman and her friends, Reitman developed his own brand of anarchism, often informed as much by his own personality as by political theory. Although he was with Goldman for ten years (with frequent, usually short, returns to Chicago and his mother) Reitman never overcame a sense of inferiority to his new comrades' intellectual abilities. "The kindest thing they can say about me is that I can sell literature," he wrote Goldman angrily in 1914. "You are a power Berkman is a force and Reitman is a *joke.*"[17]

There is little evidence that Goldman's comrades did much to welcome or encourage Reitman. On the other hand, they found much in him to criticize. For example, once he appeared for breakfast in the nude (a nudity that was more than six feet tall and often overweight and unwashed) at the home of anarchist sympathizers who were hosting them on their visit to Portland, Oregon.[18] Goldman and her comrades also despaired over his unwillingness or inability to refrain from frequent sexual encounters with countless women during all of his years with Goldman. They also criticized him on more doctrinaire grounds, challenging his commitment to or even understanding of their cause. Their—and Goldman's own—sharpest criticism of his political mettle arose from the terror and helplessness Reitman exhibited when dragged from his hotel room in San Diego in 1912. Blindfolded and hustled out of town by a band of men, he was tarred, feathered, and branded on the buttocks with the letters I.W.W. Reitman's fear and distress over the incident became emblematic to Goldman of her lover's lack of dedication to anarchist causes.[19]

In later years, both Goldman and Reitman were able to define the benefits their union held for them. Of Goldman, Reitman said, "She gave me an intellectual horizon and a soul, showed me the beauty of poetry and the grandeur of literature, and the possibility of giving worthwhile service to humanity."[20] For her part, Goldman eventually acknowledged the strength that Reitman's enthusiasm and support gave to her, despite her frequent disappointment in him. She wrote, "To be near him involved conflict and strife, daily denial of my pride. But it also meant ecstasy and renewed vigor for my work."[21] Ultimately, she told Reitman, it was the pressure of this constant emotional see-saw that finally led her to separate from him. "I never could stand uncertainties," she wrote to him in 1926. "That was the poignant part in our love. You were the most uncertain quantity I had ever loved."[22]

The Clap Doctor of Chicago

Goldman broke with Reitman when he left New York permanently to join Anna Martindale, a young British seamstress and suffragist who had frequented the anarchists' offices.[23] Martindale had moved to Chicago to await Reitman and the birth of their child. When Reitman joined her in 1917, his past connection with Goldman gave him a local notoriety that did not always make him comfortable, and shortly after his return he distanced himself publicly from his old associates. As he told a Chicago newspaper reporter, "Emma came into my life years ago and seduced me from the old world. I dropped my old ideas. I dropped Jesus Christ. But today I'm thanking God I've learned the right road—and I learned it when I came back to Jesus."[24] Whether Reitman's words were prompted by a true change of heart would be hard to say, but his past loyalties definitely complicated his search for medical offices. "Ain't you the same Ben Reitman that was traveling around with Emma Goldman?" asked the agent at the first building he tried. The scenario repeated itself over several days, until Reitman met the agent of the Bush Temple Building at Chicago Avenue and Clark Street. " 'I know you,' " Reitman reported the conversation, " 'and I know about Miss Goldman. I think we've got a place that will suit you.' And for thirty-five dollars a month he rented me a very satisfactory office."[25]

As he had in the past, Reitman sought a place near the people and activities he most enjoyed. He wrote, "From the very first day I opened my office I had a few patients and a great many friends. My office was just a few blocks from Bughouse Square, in the center of Hobohemia, near to many labor union headquarters, across the street from the Chicago Avenue Police Station, and near the County Jail. I was in the center of a modest vice area—gamblers, prostitutes and pool-halls all around me."[26] The practice was lively, untraditional, and caring, as described by a poem by Adolph Pfister, D.D.S.:

> I step into your office
> Out of the turmoil and strife and greed of the streets.
> And it is so unlike a doctor's office
> Even the wooden chairs are always awry.
> But on these chairs are live, pulsating souls—
> They love and they hate and they argue. . . .
> He understands all—he know all.
> Holding them all, idealist and prostitute, student and moron,
> conservative and radical, the believer and the scoffer.[27]

Starting or even resuming a medical practice is a slow process, and when Reitman returned to Chicago he was no longer the footloose physician that he had been ten years earlier. To augment his private practice, he turned to old contacts. William Evans, by now a former commissioner of the Chicago Board of Health, suggested that he apply for work as a smallpox vaccinator for the city. The current commissioner of health, John Dill Robertson, another of Reitman's mentors, reportedly responded to his request thus: "Ben, I'd like to give you a job, but you've got a terrible reputation. Everybody knows you're an Anarchist agitator. . . . If you'll get Doctor Evans to write me a letter, and get the police to say that they haven't got any objection to my giving you a job, I'll take a chance on you and give you a job."[28] The letter was written, and the appointment was approved. Reitman began part-time work at $5 a day, his Health Department star proudly displayed on his vest. He was assigned for several months to the African-American and hobo districts, covering them with thoroughness and success: "I worked hard and conscientiously, had a good deal of aggression, some persuasion, a little tact, and carried a considerable air of authority, so I succeeded in vaccinating a great many people."[29]

Reitman's work for the city was temporarily suspended in 1917 while Reitman, his appeals and challenges to the courts failing, left Chicago to serve a six-month prison sentence in Warrensville, Ohio, for lecturing in Ohio on methods of birth control while with Goldman. Always hounded by police and government officials for their lectures about birth control, Goldman and the other anarchists managed to escape actual prison sentences for this aspect of their work, but Reitman's medical degree may have made him a target for more severe punishment.[30] As he described it, though, the six months were not particularly punishing. As a doctor in a state prison, even though an inmate, Reitman was soon put in charge of the prison clinic, and he moved freely between the work field, field hospital, and prison clinic and hospital. He worked closely with the prison officials, at least once conducting a program for visitors in which he introduced various criminals to the outsiders. Reitman established a unique position for himself: He cared and often spoke for the outlaw or outlawed yet also worked with the official groups who tried to control the inmates' lives. As in so many avenues of his life, he bridged—perhaps, more accurately, straddled—two worlds.

Upon his return to Chicago, Reitman resumed his part-time work for the Board of Health. This time, Robertson assigned him to the Chicago House of Correction, where he directed the men's and women's venereal disease clinics for two years. Reitman soon established a familiar routine: "My venereal clinic at the Chicago House of Correction was more like a vaudeville than a clinic. We used to have more fun at the clinic, and the prisoners really enjoyed coming." The early treatment of syphilis with Salversan (606, arsphenamine, the first of the chemotherapies for syphilis) was a time of enthusiasm and optimism. As he described it:

> Although I treated thousands of men and women for syphilis and gonorrhea on a wholesale scale, with a recklessness, a daring, and a rapidity that would frighten the average V.D. specialist, I never had any fatalities, but I came very near to it many times. One day [at the Chicago House of Correction] I was shooting Salversan at the rate of "one a minute." This included the time the patient got on and off the table. Someone called my attention to the fact that one of the patients to whom I had just given an intravenous injection of neo-Salversan, had fallen to the floor. I turned around and discovered that the last five patients I had injected were laying on the floor gasping for breath. . . . We stopped for about five minutes, resuscitated the patients, and then went right on with the work.[31]

Reitman's actions defy not only recommended medical protocol but also medical credibility because Salvarsan's high toxicity made administration of the drug risky and demanded a relatively slow injection. Because Reitman refers to both Salvarsan and its next-generation drug, Neosalvarsan, which was less toxic but also less effective, however, it is difficult to know exactly what the effects and risks were. At any rate, he was obviously transgressing whatever limitations were medically recommended.[32]

Reitman had been working at the House of Correction only a few months when Robertson began a program that would become a model for other U.S. cities—the first program for systematically testing, treating, and reporting venereal disease. "There had been free V.D. clinics in Chicago that were associated with medical colleges and some charitable institutions for more than fifty years," Reitman wrote, "but I found them unsatisfactory and inadequate."[33] When Chicago opened its first municipal disease clinic on January 1, 1918, Ben Reitman was one of its two clinicians. His picture of the clinic is similar to that of his other medical ventures, with overtones of the hobo college:

> A thousand cards were printed and I distributed them in the hobo districts, and in three months we had a larger venereal disease clinic than

> the medical colleges which had been established for fifty years had. Most of the first patients at the venereal clinic were friends of mine. The clinic hours were social hours. We not only gave them Salversan, but Will Pennington used to sing to them. And Doctors and patients often sat and discussed life and love and the pursuit of woman. The clinic grew in power and usefulness and Doctor Robertson established four branches. That year the first venereal disease ordinance was passed, making venereal disease reportable, and giving the Health Department the same right to isolate, quarantine and hospitalize V.D. cases that it did small pox and other infectious disease.[34]

Besides offering this lively portrait of his practice style, Reitman also documents here the creation in 1918 of ordinances giving city officials authority to enforce treatment for syphilis.

Reitman was also present at the start of Bundesen's large-scale, sensational initiative against syphilis and gonorrhea. When Herman Bundesen replaced Robertson as commissioner of the Board of Health in 1922, he put Reitman in charge of a newly created venereal diseases clinic in the Cook County Jail. He did so, however, over loud protest from the jail's medical director, who opposed Reitman's appointment on grounds of his anarchism, prison record, and police records. William Evans and Louis Schmidt supported Reitman's appointment. "Bundesen stood by me," Reitman said, "but it wasn't easy."[35]

Bundesen's campaign began innocuously enough. On February 2, 1923, in a meeting of Bundesen, Schmidt, and Evans at the Randolph Restaurant, Bundesen agreed "to strenuously endeavor to carry out or at least inaugurate the following plan between [then] and April 1, 1923": to adopt "standards of infectivity" for gonorrhea, syphilis, and chancroid; to create a program for the forceful detention of people with infectious venereal diseases; to establish a central dispensary and clinic for the treatment of venereal diseases; and to get hospital facilities for infected people. The "standards of infectivity" had been spelled out during the preceding months by Drs. Schmidt, Evans, Reitman, and Pierce, establishing the clinical indications that would determine judgment on the health status of people with those three diseases.[36] The second goal, to develop a protocol through the city's police and court system, required only to build upon laws that had been enacted during Robertson's tenure as health commissioner. The last two goals were quickly realized in the creation of the Municipal Social Hygiene Clinic and Lawndale Hospital.

Although the agreement signed that noon at the Randolph Restaurant did not mention prevention, Bundesen's attempt to provide prophylactic kits to Chicago's houses of prostitution was obviously an offshoot of this initiative. "When Herman Bundesen was first appointed Health Commissioner of the city of Chicago, and Isaac D. Rawlings made Director of Public Health for the state of Illinois—both appointments were made by Doctor John Dill Robertson—I tried to interest them in the social aspects of venereal diseases," Reitman claimed. According to him, "Doctor Evans and I were hoping to find a way to introduce venereal prophylaxis in the houses. We thought for awhile that Bundesen and Rawlings would be as interested in venereal prevention as they were in typhoid, pneumonia and tuberculosis prevention." Reitman arranged with "Maxie, a retailer, and Sol, a wholesaler" to take the city officials on "a tour of investigation." Maxie and Sol took Bundesen, Rawlings, Evans, and Reitman to the "syndicate houses" of prostitution, where they talked to pimps and managers.[37]

Bundesen was impressed, Reitman claimed, and soon after the visit the commissioner "came out very strong for venereal prophylaxis, but when the preachers, reformers and some of the city officials came down on his neck, he put the soft pedal on prophylaxis and began to emphasize placarding the houses of prostitution, and hospitalizing infected women and men." Reitman asserted that "the men in the vice syndicates, the managers, pimps and girls all agreed to inaugurate any system of health protection and prophylaxis that we outlined to them," a statement supported at least in principle by Bundesen's interview with the *Tribune* in 1922.[38] Reitman lay blame for the failure of their undertaking on Bundesen: "But Bundesen and Rawlings followed the law of least resistance and adopted antiquated police methods. The Health Department joined the police department and the Better Government Association, the Illinois Vigilan[ce] Association, and the cheap reformers, in hounding and raiding the houses and the girls."[39] Although it is doubtful that Reitman was the sole instigator of whatever public movement toward prophylaxis was begun in the 1920s, his account of the public opposition to those efforts corroborates elements of Bonner's story. Reitman's disappointment with Bundesen suggests an early tension between the two men that helps explain Reitman's often quick impatience with Bundesen's actions in later years.

For his own part, Reitman continued to work with organized crime, at least in private, although it is hard to believe that his city employers did not know of his involvement. While he continued to examine and treat prostitutes in his city job, Reitman also worked for Jake Zuta, Al Capone's lieutenant in charge of the prostitution business. Reitman regularly tested and treated for syphilis and gonorrhea the women who worked in Capone's houses.[40] Thus, for a while Reitman worked for both the city and the mob.

Reitman's work with prostitutes was a sizable part of his practice in the early 1920s, and it continued to grow. "I feel that I know prostitutes," he stated simply.[41] The prostitutes whom Reitman treated often called him both physician and friend. He was one of the few physicians of his time to treat them with sympathy and respect. A series of letters from Margaret Parker (a.k.a. "Grace Kelly") attests to his support. Parker was sent to Lawndale Hospital for treatment of her gonorrhea in 1924. At first, she despaired of ever leaving and wrote Reitman, "Hasn't Curly [her pimp] tried to get me out? I suppose he doesn't care whether I get out or not. Gee Doctor please get me out of here it's actually awful being here." Her second letter, however, reflected a new optimism, occasioned by a letter from Reitman and an apparent encounter between him and Curly. "I hope I can go home court day and believe me I'm going to see you twice a week if I ever get gonorrhea you'll have a fine time with me. But I don't think you'll ever find another germ," she promised, closing with repeated thanks. "Well doctor please write to me, your letters surely are life savers."[42]

For all of Reitman's sympathy for prostitutes, however, at times he spoke both to and about them with unexpected harshness. For example, he wrote to Leona Clark, who had addressed a group of students on one of his "crime tours," "May I also caution you again against misrepresenting the life of the female hustler. In the talks you made at the LaSalle Street Baptist Church you told of all the fun you have and the large amount you made and you forgot to mention the disease you caught, the pinches you have, the bad influence you were to the community. The fact that you are having a good time and able to send money home to your parents in no way justifies your manner of living."[43] Reitman wrote about Clark in his reports to his bosses, often stating that neither she nor any prostitute could be trusted *not* to engage in work while infected, an unreliability he attributed to both economics and pathology.[44]

It becomes clear in reading comments such as these that Reitman, whatever his public philosophy or rhetoric, personally took prostitution to be if not immoral then at least unwise, and prostitutes to be short-sighted and selfish—and, at times, even sick.[45] Hypocritical as such a position may appear, Reitman's views were in keeping with a substantial body of "enlightened" moral and medical theory at that time. Despite the occasional harshness with which Reitman voiced those opinions, however, the sizable correspondence that he maintained with former patients makes it clear that many prostitutes turned to him confidently as someone they could trust to treat them with competence, dignity, and (usually) respect.

On one issue, however, Reitman never faltered: the prevention of venereal disease. In his autobiography, he expressed great disappointment with Bundesen in 1922, when the commissioner abandoned prophylactic kits for raids and quarantine. "Morals courts, regardless of their merits and regardless of the tactics of the police or other drastic methods of combatting vice, are not halting the spread of venereal disease," Reitman eventually concluded. "Preaching and force are equally futile. The sex urge cannot be controlled, whether the individual is an anti-social vicious pimp or cultured thoughtful gentleman with a marked social conscience. There's only one hope, and that is prophylaxis prevention."[46]

On this issue, Reitman was in direct conflict with Surgeon General Thomas Parran, whose scientific approach to syphilis nevertheless harbored his personal belief in the necessity—and superiority—of sexual restraint. When Parran spoke about the prevention of syphilis, he spoke primarily in terms of moral prophylaxis, advocating it above "mechanical protection"—the condom—about which he had little to say, or chemicals that could attack the spirochete, which he considered appropriate only for "emergency . . . disinfection."[47] Although the Public Health Service was engaged in some experiments around chemical means of prophylaxis, at least in the early years of its antisyphilis campaign, direct questions from Chicago about chemical prophylaxis received vague answers.[48] In February 1937, Assistant Surgeon General R. A. Vonderlehr quoted a memo from an assistant, a Miss Martin, when he responded to a query from Dr. Oliver C. Wenger, the Public Health Services special consultant in syphilis control for the Chicago area: "In the studies of comparative effectiveness made in this country the names of the tubes [of chemicals in various prophylactic kits] are not given, nor is there anything to indicate that they were commercial preparations. . . . In a few of the abstracts from foreign literature names of proprietary preparations are given, but these would not be sold in this country."[49] In general, the answer, which the Public Health Service would give repeatedly over the next few years, was that there was not a chemical prophylactic as yet considered effective enough to be acknowledged by the government, a position that allowed it, at least in its own view, to abstain from further discussion of chemical means of preventing syphilis.[50]

As an employee of the Chicago Syphilis Control Program, Ben Reitman was forbidden to speak about the prevention of syphilis by any measures other than moral. Although it is hard to imagine Reitman agreeing to such a dictum, it is even more startling to see the following item in his job description: "To propagate venereal prevention and prophylaxis." That seeming violation of Parran's orders, however, is partially explained by the last item in the description: "To keep in the background and see that all reports go directly either to the President of the Board of Health or to the Chief of the Social Hygiene Division."[51]

A careful distinction was created—and rigidly maintained. In 1937, Reitman created his own organization, the Chicago Society for the Prevention of Venereal Disease, a "paper organization, a 'one man show,' " headquartered in his medical office, now at Room 817, 32 North State Street, funded totally from his own pocket.[52] In his talks "around town" Reitman only identified himself as an employee of the Social Security Board of the Board of Health to "individuals who [were] already aware of this association and from whom he [was] seeking data."[53] Whenever he spoke about preventing disease, however, he would introduce himself as director of the Chicago Society for the Prevention of Venereal Disease. To monitor Reitman's compliance, his bosses would read the regular reports he was required to make about his comings and goings.

Thus, Reitman could carry the message of prophylaxis without technically violating the federal injunction against doing so. He did not confine his propagandizing to Chicago, however, and by late May he was writing directly to Surgeon General Parran and Assistant Surgeon General Vonderlehr over his signature as director of the Chicago Society for the Prevention of Venereal Disease.[54] These letters were duly copied into his Venereal Disease Control Reports. Six months later, he warned Reuben Kahn, "If you find a long-haired, articulate heckler bothering you in Indianapolis at the meeting of the American Association for the Advancement of Science, don't be surprised."[55]

The issue of prophylaxis will receive further consideration in subsequent chapters of this book. It is important to note here, however, that Parran's public policy toward the disease—a disease that, he insisted, should be discussed openly—nevertheless remained remarkably silent about the range of possibilities for prevention. Reitman's insistence on the naturalness of sexual behavior necessarily led him to desire ways to accommodate the prevention of syphilis to that behavior. By contrast, Parran's appeal for moral prophylaxis passes judgment on sexuality itself.

Ben Reitman as doctor to U.S. railroad crew working in Mexico (circa 1904). (From the collection of Ruth and Joel Surgal)

Caricature of Reitman from unidentified mural, with characteristic hat, cane, and tie—as well as his book, The Second Oldest Profession, *and syringe with which he is presumably giving treatment for syphilis (circa 1935). (From the collection of Ruth and Joel Surgal)*

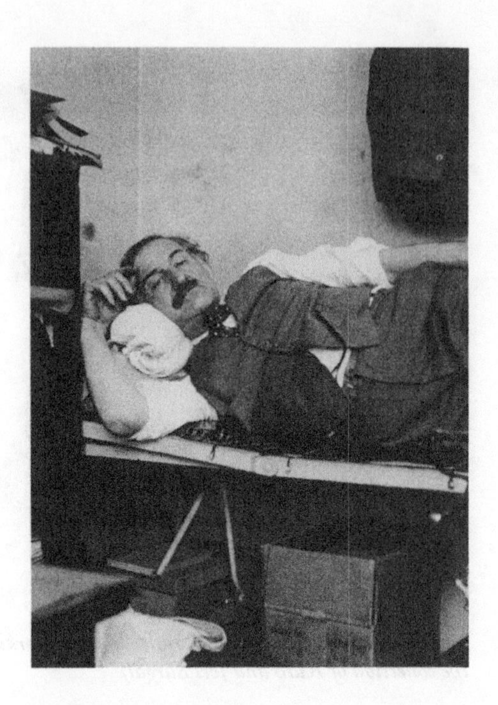

Reitman reposing in his office (circa 1937). (From the collection of Ruth and Joel Surgal)

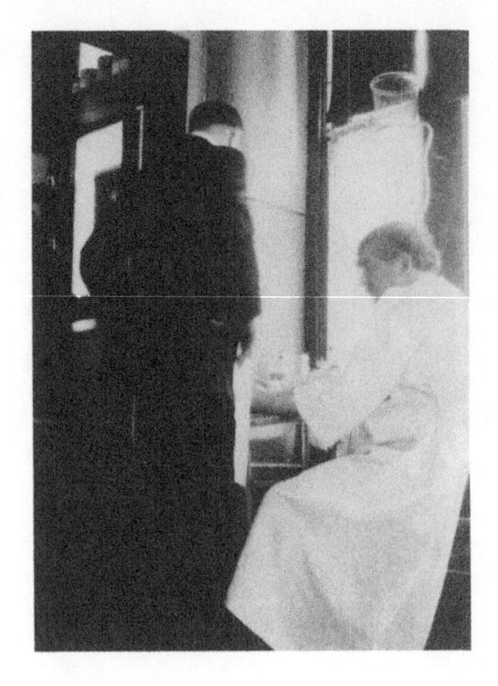

Reitman examining a patient (circa 1937). (From the collection of Ruth and Joel Surgal)

Reitman dictating to his stenographer (circa 1937). (From the collection of Ruth and Joel Surgal)

Emma Goldman, anarchist leader and lecturer. (From the collection of Ruth and Joel Surgal)

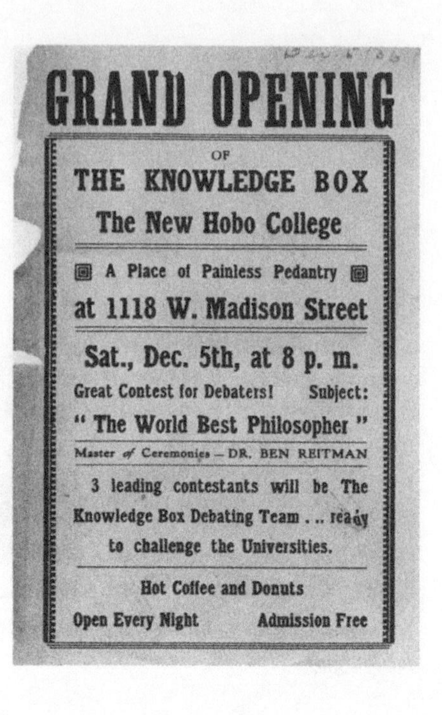
GRAND OPENING

OF

THE KNOWLEDGE BOX

The New Hobo College

A Place of Painless Pedantry

at 1118 W. Madison Street

Sat., Dec. 5th, at 8 p. m.

Great Contest for Debaters! Subject:

" The World Best Philosopher "

Master of Ceremonies — DR. BEN REITMAN

3 leading contestants will be The Knowledge Box Debating Team . . . ready to challenge the Universities.

Hot Coffee and Donuts

Open Every Night Admission Free

Flyer for hobo college debating team, which regularly went up against (and regularly defeated) teams from Northwestern University and the University of Chicago (1936). (From the collection of Ruth and Joel Surgal)

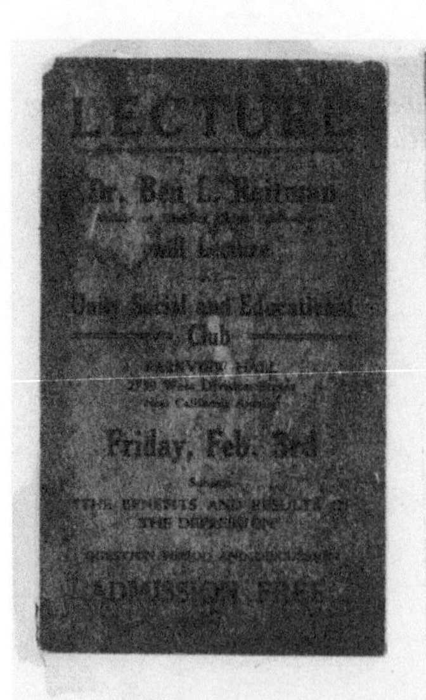

PHENIX LODGE No. 1141, A.F. & A.M. 32 W. Randolph St. Hall A-3 Chicago

SENSATIONAL DEBATE!

LOVE OR JUSTICE

WHICH IS THE MORE EFFECTIVE FOR HUMAN PROGRESS?

Dr. BEN L. REITMAN (The Billy Sunday of Chicago)
Well known Lecturer and Author
Will speak on "LOVE" (Ladies Take Notice)

H. B. RITMAN Dynamic Orator
(Former Assistant States Attorney)
Will speak on "JUSTICE" (How about it Men!)

Monday Eve., January 28th, at 8:30 o'clock

LADIES BRING YOUR HUSBAND AND FRIENDS

ELI EPSTEIN, Worshipful Master — Lodge Opens 7:30 P. M. Officers Inst. 5:30 P. M. — M. M. KRAMER, Secretary 1120 N. Oakley Blvd.

THE SEVEN ARTS CLUB

Invites you to bring your friends to the Gala Banquet and Lecture Forum to be given at THE GREEN SHUTTERS, 212 East Ohio St., (just east St. Clair Hotel)
THIS SATURDAY NITE, OCTOBER 19th.

This is Celebrity Nite in Bohemia. Come and meet them at our Banquet, from 7 to 8 P. M.. Splendid 5-course home-cooked dinner will be served (only 50c) At 8:30, following the Peacock Alley Chorus, a-la-boheme, the noted Sociologist, Litterateur and Physician

DR. BEN L. REITMAN

whose amazing series of Underworld Revelations is now running in the "Real America" Magazine, will lecture "SEXUAL SECURITY, OR THE PROFITS OF VICE"—Does 'Easy Virtue' bring Easy Money—Does the Easiest Way Pay? Many prominent social workers, criminologists, medical and legal personages will be present and participate in discussion.

GENERAL JACOB S. COXEY

of "Coxey's Army" fame, the world-famous figure of his age, a veteran orator of eloquence and power "IF I WERE PRESIDENT—HOW I WOULD SAVE THE NATION." Hear, see and meet Gen. Coxey at our Gala Banquet, Don't miss it! The Great, Paramount, Burning Issue of these times:

"CAN DEMOCRACY SURVIVE THE DICTATORSHIP ERA?"
MONTY RANDALL vs. A. BUCKLAND PLUMMER
EDW. CLASBY JESSE QUITMAN

A Nite of Stars! A Constellation of Personalities! In Dynamic Debate!
Please Phone Dinner Reservations: (Kenwood 7151).

A sampler of Reitman's lecture engagements (circa 1936). (From the collection of Ruth and Joel Surgal)

Starry-eyed Dinosaur

In seeing himself as medical and moral advisor of both his patients and the officials for whom he worked, Reitman was setting himself an ambitious goal. Furthermore, by 1937 Reitman was playing to a public that no longer saw him as larger than life. At fifty-eight, he was not an old man, but he was clearly worn by years of full, hard living. His health was failing. He was overweight, diabetic, and had a bad heart, with suggestion of liver problems. His medical practice, although busy, had not completely sustained him for several years, and now he had more need than ever for money, because he had recently become the father of a new daughter, Mecca, born in 1936. Three more daughters would be born of Reitman's union, his fourth and last, with young Medina Oliver. Oliver, a graduate student when she met Reitman at a lecture he was giving at Maxwell Street Clinic, worked as a laboratory technician after the births of their children and provided most of the family's income. Reitman's work with the Chicago Syphilis Control Program was, then, important to him not only for the nature of the work but also for the income it provided.

Fortuitously, he was working with a problem dear to his heart. Still, it is hard to imagine anyone keeping up—cheerfully—with the workday he described in the following letter: "We, too, are busy, running the Chicago Society for the Prevention of Venereal Disease, the Hobo College, the movement to get jobs for men in jail who are waiting to be paroled, leading a committee who is collecting old radical, free love, birth control, labor, anarchist, and radical literature from the period of 1865 to date. The fact that we have a large office practice, make some calls for the medical relief, and are on the payroll of the Social Security Board (and must give them 7 hours a day to earn our salary) does not interfere with the fact that we have plenty of time to read."[56]

Reitman's seven-hour days for the Social Security Administration, in addition to "propogat[ing] venereal prevention and prophylaxis" and "keep[ing] in the background," included sixteen other specified activities that embraced reading and reporting on all new books and articles on syphilis and gonorrhea; lecturing to "underworld, hobo and other groups," boys' clubs, and "open-air [soapbox] settings"; visiting all sites of activities in Chicago related to the testing and treatment of venereal diseases; investigating conditions and preparing statistics about venereal diseases within the worlds of prostitutes and the transient or homeless; checking all statistics about the program that were released to the public; and, very generally and vaguely stated, "populariz[ing] the activities" of the program.[57]

It was a prodigious assignment; moreover, it was to be documented by the daily written reports to his superiors. But no task seemed impossible to Reitman. "The best job in the world," he wrote to his bosses about his work as an investigator for the Social Security Board, a job made possible by the new funding to states from the Social Security Administration.[58] To his old friend and mentor William Evans, now retired and living in Aberdeen, Mississippi, Reitman wrote, "Dr. Bundesen, Taylor and Schmidt have given me considerable leeway and a good deal of encouragement. They have stood gracefully for my criticism and were only provoked when I adopted a belligerent attitude toward Rachel[le] Yarros." Choosing not to elaborate on this "attitude," he went on to describe his new work: "I am not working in the clinic as a physician but am employed as an investigator gathering data and making reports. We studied the question of free clinics in Chicago; the amount of blindness; congenital syphilis; the number of syphili[tics] of the insane in our Cook County institutions; the amount of venereal disease amongst the prostitutes; the need of a south side branch venereal disease statistics." Only one thing was lacking: "I have been impatient because the department would not let me come out boldly for prophylaxis and the distribution of the family prophylactic venereal disease campaign, but they tell me to hold my horses." A month earlier, Reitman told Evans, he had complained to his bosses about the strain this silence caused him. His next words to Evans may well reflect the response they made to him: "ACTING UPON ORDERS, WE HAVE NOT SPOKEN AS A REPRESENTATIVE OF THE BOARD OF HEALTH—ALWAYS AS A DIRECTOR OF THE CHICAGO SOCIETY OF VENEREAL PROPHYLAXIS." Then he added, "It is highly preferable that we speak 'as one who has authority,' but until we have orders to do so, we will go on as a lone wolf."[59]

Such a situation may have both accommodated Reitman and allowed his bosses an outlet for information they, too, believed should be a part of the program. Nevertheless, such a strategy seems destined from the start to exacerbate any existing tensions between Chicago and Washington on this issue. Moreover, torn between his desires to play an official role in syphilis control with people for whom he had worked in a happy earlier time, his growing need for steady work, and his own convictions about the treatment and prevention of syphilis, Reitman placed himself in an impossible position from the start. For all of his enthusiasm for his new job, most of his writing at this time reflects the tenuousness of his position, the sense that he was with the Chicago Syphilis Control Program on their sufferance, watched and cautioned by his supervisors and old friends.

Was he seen as a loose cannon, best guarded by being brought into the group but kept close track of? Were his association with and knowledge of certain groups of Chicago's population valued, with tasks assigned him that would capitalize on his expertise? Or was he so entrenched in the system and history of Chicago public health politics that he would always be there, at times needed but at other times merely tolerated? To some extent, all of these suppositions are probably correct. Reitman's own vision of himself and his abilities at this time in his life often faltered. In 1938 he wrote:

> If there were only someone to write to
> Some[one] who I though[t] cared [if] I was tired and weary
> Because I had a weary day
> To feed a dozen hungry souls. . . .
> This is what I have been doing for ten long hours
> They took all of my money, most of my strength nearly all of
>   my hope
> As I homeward wend my way I have nothing to take with me
> Just a weary broken man
> And there ain't anybody to write to.[60]

Thus, for all his bravado about being part of an important team, Reitman realized that he was not as central to that team as he would like to be. In an early letter to Lawrence Linck, the Public Health Service's appointed director of the Chicago Syphilis Control Project, he painted himself in a very inferior light:

> I just want you to know that I have been thinking about syphilis for 40 long years, and when I see you bright, young men and women, with your splendid preparation for mathematical and statistical studies, and your aptitude for research work, and your capacity to think in astronomical terms, I feel very much cheated that I did not have such fine training to begin such an important study. . . . May I say in passing that you're practically the only group who take me seriously or pay any attention to me; because of my boisterousness and lack of tact, more than two-thirds of the syphilographers and syphilis organizations and health officials have dropped me like a hot potato. But merrily we roll along, joyfully.[61]

Besides being kept in a marginal position in his medical work, Reitman by 1937 was also no longer a part of other mainstreams. Although a member of the left-wing, prestigious League of American Writers (which included such authors as Nelson Algren, Sherwood Anderson, VanWyck Brooks, Langston Hughes, Upton Sinclair, Meridel LeSueur, and Paul de Kruif), Reitman's love of the vulgar and outlandish had made him more of a caricature than a character in Chicago literary circles.[62] One Chicago man recalled, "Whenever I saw Reitman, he seemed to be the butt of bad-taste jokes by soapboxers who, espying him in the crowd, seemed impelled to show off their acerbic commentary at his expense; and Reitman, evidently to avoid confrontation, from which he seemed by temperament to shrink, would try to disarm the jokester and his barbs by joining in the laughter and even adding a comment or two at his own expense, to show he relished the joke. . . . One only regarded him curiously as the man who had once been Emma Goldman's lover."[63]

The sense that Reitman was an anachronism is vividly conveyed by Elmer Gertz and Bruce Milton in their unpublished biography of Reitman, written during the years of the Chicago Syphilis Control Program. "For sixty years," they wrote, "Ben Reitman has foraged and ranged, like some prehistoric nightmare, through the green pastures of a too often promised land. And now, bogged down by time and countless conflicting motives and utterances of his past, he lies trapped, a helpless and ludicrously tragic figure, snared in the slime and muck of social disuse, a dinosaur with starlight in his tired eyes, dimly dreaming of past combats, of mighty bellowings, of his thunderous progress through the jungle, bowling over all that impeded a joyous and violent journey, driven by hungers that he could never define nor gratify."[64] Whether this image was ever true, during the months that he worked with the Chicago Syphilis Control Program, Reitman was able to enter in a "combat" that was for him both "joyous and violent" but that left him, once again, with a "hunger that he could never . . . gratify."

Ben Reitman not only reveled in hyperbole, but he also seemed to elicit it from others. He wanted the world to see him as a crusader, a radical, the intimate of important and notorious people, and himself important and unique as well. If not a Lancelot, he would at least like to be a Quixote, a prophet, and—if need be—a martyr. Whoever first had the idea for the earliest venereal disease programs in jails or houses of prostitution, Reitman was a part of those programs from the start, and his knowledge of the players and their politics was intimate and longstanding. Moreover, where Herman Bundesen conceded to public (and perhaps political) pressure in

1922, Reitman never recanted his views of syphilis control. Which man was wiser, either in the moment or the long run, could be debated. Certainly, Bundesen's influence carried more weight in 1937 than did Reitman's—but how does one weight the cost of political, medical, or perhaps even moral compromise?

Ben Reitman was an egotistical man who wanted and needed the attention and admiration of friends, family, and public. He was bright, with a quick but probably not a deep mind. He was passionate and demonstrative of those passions, whether love or hatred. He would have been a difficult man for anyone to have as a friend, yet many of his closest friends found Reitman's fierceness to be his greatest quality. The intensity of his impatience with hypocrisy and half-way measures sounds in every word of his writings, whether official or personal. It is this fierceness of purpose and belief that Emma Goldman found to be Reitman's most valuable quality. It is this quality that he demonstrated over and over again in his crusade against venereal disease, in demonstrations that different interpreters will judge differently—as successful, reasonable, impetuous, or impossible.

CHAPTER THREE

Taking the Plunge

Legislative change provided the second front of activity for the Chicago Syphilis Control Program in its first summer of operation. It was heralded by passage of the Saltiel Law, which went into effect July 1. The Saltiel Marriage Law made Illinois the fourth state in the nation (after Oregon, Montana, and Connecticut) to require testing for venereal disease before the issuance of a marriage license.[1] In 1937, nearly half of the states in the United States had laws requiring a physician's assurance that one (always the man) or both parties applying for a marriage license be free of venereal infection. Most of these laws, however, did not require laboratory examination. Moreover, enforcement had become increasingly lax.[2]

The Illinois law was different from these earlier laws in several important ways. Although called the Saltiel Law, it was actually a combination of laws. The Saltiel portion, introduced by State Representative Edward Saltiel and State Senator Harold G. Ward, passed resoundingly, with a vote of 105 to 4 in the house and 31 to 1 in the senate.[3] It required that couples married in Illinois certify their health by laboratory examination within fifteen days of their application for a marriage license. Infected persons would not be issued the license until they could document that they were in treatment. Moreover, a license granted to a couple could be used for only thirty days after it was issued. Reapplication would require recertification of health.

The second bill passed into law at this sitting of the legislature was introduced by a legislator named Graham. It instituted a mandatory three-day waiting period between the application for and issuance of the marriage license. This bill, *Tribune* reporter Percy Ward remarked, was unique and made Illinois the first state to enact a law deliberately aimed at stopping gin marriages.[4] *Gin marriage* was the term popularly applied to weddings that occurred, usually, without premeditation and after an evening of drinking alcohol. Thus, although the Saltiel Law was passed in the name of public health, the Graham Law made no bones about its moral condemnation of the combination of liquor and marriage. By such a juxtaposition of laws, premarital legislation in Illinois from the start linked syphilis in many people's minds with—however mistakenly—lascivious, irresponsible behavior.

"A Good Law"

The results of the new laws, always referred to jointly in the *Tribune* as the Saltiel Law, were immediate. "Marriage Test Bill Signed; In Force on July 1" the June 24 headline read innocuously enough, but the seriousness of the law was spelled out in the severity of its consequences: a $100 fine or six-month jail term for any county clerk issuing a license without receiving a certificate to verify testing, the same fee and term for physicians or laboratory technicians who falsified certificates, and a similar fine or three-month sentence for any applicant who violated the law in any way.[5] The next day, support for the new law echoed more resoundingly, with the headline, "Hail New Law as Powerful Aid to Syphilis War," introducing the article that included, at some length, the history of Chicago's fight against syphilis, beginning with laws as far back as 1911, Chicago's open discussion of syphilis after World War I, and finally the new objectivity of epidemiology that would now make syphilis once more an unabashed, household word.[6] On the next page of the *Tribune* a much shorter article appeared, its headline in smaller type: "Seeks Opinion on Sections of Marriage Law." The section of the law in question was the requirement that medical doctors test marriage applicants, a restriction that, probably unintentionally, prohibited doctors of osteopathy from performing the examination. At the end of the article, an unrelated comment noted that Chicago officials were expecting a record number of applications for wedding licenses during June in direct response to the new laws.[7] This aside proved to be a momentous understatement.

Two days later, on Sunday, June 27, headlines in the *Tribune* read, "Rush Marriage Bureau to Beat Hygienic Law." A photograph showed a sea of heads and shoulders massed before a counter of busy clerks as they labored to issue 793 licenses in eight hours.[8] Another two days later, the *Tribune* reported, "1,119 Couples Get Licenses in Day to Beat Marriage Law." This number fell five licenses short of the record set on April 9, 1917, the day before the draft of soldiers to fight in World War I went into effect. The Marriage Bureau extended its working day an extra hour to accommodate the surge.[9] Working hours were extended even further and the old record broken on the last day of June, as 1,407 licenses were issued in eleven hours. "Jam Marriage Bureau to Beat Test Deadline," the front-page headline read that day. The total number of marriage licenses issued during the last week of June was 6,266.[10]

Many of the couples applying for a license cited as their motivation the nuisance and expense of the examination or the preferability of having a license that would be valid beyond thirty days.[11] Although none of the couples who were interviewed at the Marriage Bureau challenged the importance of the test itself, they complained of its expense, even though previous *Tribune* stories had made it clear that city and state health departments' laboratories would administer it at no charge. Even as couples were crowding the bureau in record numbers during the last week of June, they maintained that the new law was not influencing their actions. "I just talked to a fine young couple," said County Clerk Michael J. Flynn. "They said they'd been going together six years. They had a date tonight, were out riding, and just decided they'd get a license."[12]

The passage of a law had become a media event. Whatever people's stated reasons for receiving a marriage license before July first, the *Tribune*'s increasingly sensational coverage of the crowds in the bureau sent mixed messages to its readers. Headlines that linked the rush to obtain a license to a desire to "Beat Hygienic Law" or "Beat Test Deadline" suggested that the test itself was undesirable, a connection that was never borne out by the fuller story printed in the paper. Repeated stories about attempts to "beat" a law create negative images of not only the law or the test that it mandated but also of those men and women who would evade that law. Such visible desire of so many people to avoid a test for syphilis seems on the surface to contradict the *Tribune*'s insistence that syphilis was "not a badge of disgrace or immorality."[13]

The *Tribune*'s coverage of the subsequent, even more sensational events in the wake of the Saltiel Law further demonstrates the complexity of telling the news. Again, the situation began unremarkably. On the first Sunday in July, the paper reported the appearance at the Marriage Bureau of the first couple to be denied a license under the new laws. They applied for the license on Saturday, and when asked for their examination certificate, they said that they hadn't heard of the new requirement. "Marriage Law Is

News; Couple Denied License" read the headline, and the couple was depicted as more comical—had they not been reading the front page of the *Trib* for the past week?—than criminal. The story carried further interviews with other couples at the Marriage Bureau who knew about the new laws, intended to comply with them, and attested to their wisdom. The close of the article foreshadowed with entertaining understatement (from a contemporary viewpoint) the events to follow: Thirteen couples received marriage licenses that Saturday and had been immediately married, *not* in Chicago, but across the border in Crown Point, Indiana.[14]

The boom at Chicago's Marriage Bureau was followed after July 1 by a bust of equally astounding proportions. After a record 9,925 licenses were issued in June, only 619 couples took out licenses in July, and the number in subsequent months continued to run 30 to 50 percent below the averages of the previous year.[15] In the weeks following the installation of the new marriage application laws, many couples who turned away from the Chicago marriage bureau traveled instead to neighboring states that had no similar laws, particularly Indiana, Michigan, Wisconsin, and Iowa. Indiana received the lion's share of the traffic, and marriage bureaus in border towns such as Crown Point and Valparaiso were soon doing a land-office business. The gin marriages in Illinois that legislators had intended to squelch were now, so it appeared, taking place elsewhere.

Subsequent news stories focused not on Saltiel's bill but rather on Graham's legislation of the three-day wait after application before an Illinois license was issued. The target for this law, marriages assumed to be inspired at least in part by inebriation, was made clear a week later in two stories in the *Tribune*, one hailing Illinois for being the first state to enact a "gin marriage law" ("New Act Puts Curb on Gin Weddings") and the other announcing, "Gin Marriage Law Blocks Forty Couples at License Bureau."[16] The second article went on to state that seventeen of these couples immediately proceeded to follow the new protocol without apparent objection, but the implication of the headline is clearly accusatory, hinting at questionable motives for marriages that were "blocked," thanks to the vigilance of the new state law.

The articles soon became even more damning. *Tribune* reporters visited the marriage license bureaus and justices of the peace that lined whole streets in Crown Point, interviewing Chicago couples. They particularly delighted in noting the ages (the younger the women and the older the men, the greater appears the likelihood that their stories would be told) and degrees of inebriation of the couples, and the idiosyncrasies of dress or speech of the justices of the peace. On July 12, Marcia Winn reported that 262 of the 310 licenses issued in Crown Point since July first had been purchased by Chicagoans. (Although we do not know what the numbers or ratios had been previously, the newspapers assured readers that this was a dramatic increase in both.) "Marriage Law Fugitives Flock to Crown Point," her headline read, the connection with lawlessness pointedly made. Here, though, lawlessness was charged against people avoiding the syphilis test as well. "It isn't that we mind that medical examination," one man was quoted as saying, "but it costs a lot, and I just make $25 a week. Then, too, we didn't want to wait." The "marrying justice" told Winn of another couple who had applied for a license in Chicago but went to Indiana when their test results did not arrive at the end of the three days.[17] One attempt to answer these charges of costliness appeared on August 1, after reporter Doris Lockerman spent an evening interviewing couples at Crown Point. She wrote, "These couples seem surprised when informed that they might both have had their examination for $5 and their license for $3 in Illinois. This is considerably less than many couples pay here [Crown Point, Indiana] into the organized business that sometimes raises marriage license rates as high as $10."[18]

At one point during the operation of the Chicago Syphilis Control Program, Dr. Rachelle Yarros complained that the public had not been educated well enough about syphilis and the new laws before the program began, being especially uninformed of the options for obtaining the tests and treatment.[19] Certainly, the couples who appeared at Chicago's Marriage Bureau or traveled to Crown Point often revealed their misunderstanding about the circumstances of the syphilis test. The *Tribune*, for all its pride in helping support the work of the Chicago Syphilis Control Program, contributed to that confusion. Undercutting Winn's careful explanation of the actual cost of an Illinois marriage license compared to one in Crown Point, for example, is that article's sensational headline, "Dodge Marriage Law and Laugh in Crown Point." Similarly, another story reported on a couple who tossed a coin to decide if they would spend their money on a wedding or alcohol: They left the bureau in search of a drink. The journalist concluded her story by observing that "none of the couples from Illinois showed in their manner any embarrassment at the thought that others might believe they were purposely evading the *health* test."[20]

Even though some avoidance of the Saltiel Law was ostensibly grounded in economic reasons, the notoriety around both marriage laws continued to intensify. Stories about the various Indiana businesses appeared almost daily in the *Tribune*, with the longest, liveliest stories appearing after the weekends, when business was particularly brisk and colorful. An equal source of outrage was to be found in the county clerks who profited from the profligates. For a while, the *Tribune* abounded with stories and pictures of the marrying justices of Crown Point and Valparaiso, whose offices and billboards lined the streets and who advertised their services on matchbooks given away in the licensing bureaus. One reporter calculated that county clerks in the Indiana counties neighboring Chicago would average, conservatively, $60,000 above their annual state salary if the current rate of business continued—a princely sum, as it was pointed out that President Roosevelt's salary was $75,000.[21] The greatest uproar among journalists, lawmakers, health officials, and clergy was occasioned by "marrying justice" George W. Sweigert, clerk of Lake County, Indiana, whose income from his fees had "been estimated at $75,000 annually over and above [his] $4,800 salary."[22] Anger against Sweigert and his colleagues mounted as premarital testing laws were passed in Wisconsin and Michigan, swelling his coffers even more. Iowa's clergy and state officials joined in a pledge not to marry or grant licenses to drunken couples from Illinois.[23]

When stories of inebriated couples and teenage brides lost their novelty, the *Tribune* turned to accounts of regret and deception. The first story appeared on August 14, with an article entitled "Link Three Broken Marriages to Indiana Justice." It told of two drunken women who were threatened into marriage by their companions. Four days later, the *Tribune* told the tales of two other Crown Point couples who filed for divorce in Chicago courts. The next week, in "Fifth Crown Point Wedding Strikes Snag in a Week," readers learned about a man, not only married but also with a pregnant wife, who used a false name to marry another woman. A week later, the *Tribune* attributed four more divorce petitions to Indiana weddings.[24] Such stories continued throughout the next several months.

As the petitions for divorce or annulment came to court, Illinois judges listened without sympathy. They often handed down decisions similar to the one delivered in the first such case brought before Superior Court Justice, John L. Lupe. The couple, married on August 5, appeared before Lupe on September 30. The *Tribune* covered the story in detail. "The case, the first of its kind in Cook County courts, was filed by Mrs. Maria Hughes Steinke, 35 years old, 1439 South Cicero Avenue, Cicero. She testified that she and Edward Steinke, 40 years old, had driven to Gretna Green, Indiana, to avoid complying with the Illinois laws." The Steinkes were not the first couple from Illinois that had ever petitioned the courts to dissolve a marriage made in Indiana. This time, though, the plaintiffs were invoking the very laws that they had sought to evade in the first place in their attempt, one month later, to nullify the marriage—a titillatingly circular argument. Judge Lupe "ruled that a deliberate evasion of Illinois' new hygienic marriage law does not constitute grounds for the severance of wedding bonds contracted elsewhere," and he remonstrated the petitioners, "The law with which you refused to comply is a good law, designed to protect the children of future generations."[25] Again, whether the Steinkes were evading the syphilis test, the three-day wait, or both, Lupe refers only to the "hygienic law" as the motive for the couple's consequent lawlessness. Apparently, Cook County judges had determined that the just desserts of those who married in haste, was, truly, to repent at leisure.

The Chicago Tribune *and Syphilis*

In covering the story of the Chicago Syphilis Control Program, the *Chicago Tribune* cast itself in a major role. "Tribune Pioneer in Open Fight on Syphilis Plague" read a headline written by Stanley Armstrong (who was among the paper's top political writers and eventually became a day city editor).[26] The paper proclaimed its "frankness" compared to other newspapers of the city, claiming to be "First to Break Venereal Taboo" by having printed the word *syphilis* in 1921. "In the following year," Armstrong recounted, "this newspaper began its long campaign to bring venereal disease into the open."[27]

Other writers for the *Tribune* also took pride in the newspaper's past efforts to speak the unspeakable. Arthur Evans (another top writer of the 1930s and a major political reporter during the 1940s, whose name appeared almost as often as Armstrong's in bylines on antisyphilis stories) reported that Chicago had been chosen as the first to undertake a city-wide referendum for venereal disease testing "because of the lead taken by THE TRIBUNE in its long campaign to bring the scourge of venereal disease into the open where it can be fought."[28] One editor boasted that the newspaper's publicity campaign "has put the salt in the lead of this great movement. The driving force is frank and full publicity. . . . The TRIBUNE led in breaking this taboo down in an attack on quacks in 1913."[29] The *Tribune* also quoted others who sang its praises, such as Secretary of the Board of Health Louis Schmidt and Barnet Hodes, corporation counsel for the city of Chicago, who "credit[ed] THE CHICAGO TRIBUNE with the publicity campaign that has given the fight on syphilis its present impetus."[30]

There is no evidence to suggest that the *Tribune* was selected for or acted as a mouthpiece for the program. On the contrary, there is some indication that Chicago officials might originally have encouraged a broader support of its work than it eventually received. For one thing, the editor of the *Chicago Daily Times*, Richard Finnegan, and not a representative of the *Tribune*, spoke at Mayor Kelly's kickoff in January. Finnegan's words were, however, cautious. "If you create news about syphilis, you will force the newspapers to print it," he opened his comments, and he closed with another sentence to similar effect: "If you do those things which are really news, the newspapers will not neglect it."[31] The words sound more cautionary than supportive because they seem to challenge the Chicago Syphilis Control Program to do work deserving coverage, a proposition that refuses to accede to the bidding of the Board of Health.

The *Tribune*, however, threw itself early and wholeheartedly into the fight. "Special mention must be made of the Chicago Tribune, which took the lead in the educational program," O. C. Wenger reported. "This great newspaper, with a circulation of 800,000 daily, courageously broke the taboo of the press and printed frank, straight-forward facts with regard to the venereal diseases. The paper assigned special writers to cover the venereal disease control program and has steadfastly supported the entire campaign. The value of the Chicago Tribune to the program cannot be over estimated."[32] Assessing the extent of a newspaper's sway is obviously a precarious undertaking, and picturing the *Tribune* hand-in-glove with City Hall in controlling the public's knowledge about and image of the Chicago Syphilis Control Program would be a misrepresentation of both the relationship of the city to the *Tribune* and the power of either of them. Nevertheless, the work of the program fit, more than it did with any other Chicago newspaper, within the *Tribune*'s tradition and mission.

Newspapers *do* influence public debate and demands, but, in turn, its audience also shapes a newspaper, as not so felicitously observed by Alexander Wolcott, who commented that "Germany was the cause of Hitler just as much as Chicago is responsible for the Chicago *Tribune*."[33] Such a charge could probably have been leveled at some of the other newspapers competing with the *Tribune*. In 1937, it was only one of many newspapers in Chicago and certainly not the only one to print stories about the Chicago Syphilis Control Program. In fact, other papers, such as the African-American *Defender* and the socialist *Daily News*, printed stories that were, respectively, more racially and economically sensitive to the policies and activities of the program than some of the *Tribune*'s coverage.[34] A comparison of the coverage of the Chicago Syphilis Control Program by the various Chicago newspapers, however, will not be undertaken here. More relevant to this study is an overview of the *Tribune*'s policies and practices before 1937.

When Joseph Medill and physician-turned-journalist Charles Ray bought the *Chicago Daily Tribune* in 1853, it had already been in publication for more than five years. As the city's political voice for the Know-Nothing party and temperance groups, it competed with the *Daily Democrat* and the Whig-oriented *Daily Journal*. Although its new owners soon shifted the paper's politics away from the ultra-conservatism of its former owners, the *Tribune* was always to be politically cautious. For example, the newspaper generally supported labor issues because of its hatred of business trusts and monopolies, yet it remained uneasy of feared communist proclivities of labor movements. It advocated temperance, yet condemned the violence and corruption resulting from Prohibition, and, later, it appreciated the jobs provided by and resulting achievements of such agencies as the Works Progress Administration while remaining relentlessly opposed to the New Deal.[35]

Early in its career, the *Tribune* addressed itself to issues other than local and national politics. *Tribune* historian Lloyd Wendt notes, "*Tribune* policy changed from a political orientation to one primarily directing readers to lead better moral and religious lives." It printed news from and about local churches and temperance organizations. James Keeley, managing editor in the early years of the twentieth century and for a time the honorary dean of journalism at the University of Notre Dame, urged journalists to adopt a philosophy of service: "The real newspaper has outgrown the looking-glass stage as the sole object of its existence. . . . The big development of the modern newspaper will be along lines of personal service. The newspaper that not only informs and instructs its readers but is of service is the one that commands attention, gets circulation, and also holds its readers after it gets them."[36] Keeley acted on this philosophy early in the second decade of the 1900s by initiating regular columns on beauty care and charm, humor and verse, readers' complaints about public facilities or services, advice to the lovelorn, and health information. "How to Keep Well," the first column of its kind in the country and quickly syndicated, was written by no less a personage than Chicago's health commissioner, William A. Evans. The *Tribune* also undertook seasonal or special service campaigns, such as sending baskets to the needy at Christmas and opposing the sale of fireworks for July 4 celebrations.

Of a different tenor, but still considered part of its mission of service, Keeley sent numerous reporters to reputed physicians' offices in 1913. Complaining of nonexistent aches and malaise, these reporters hoped to expose and put out of business the "quack doctors" operating in and around Chicago.[37] Although no name appeared on the resulting articles, *Tribune* writer Armstrong reported in 1937 that they had been written by Dr. Evans himself.[38] The series, which ran throughout November and December of 1913, spoke often and directly about syphilis. Its first article appeared on October 27: "Cure-Fakers Find Disease in Well Men." The story began by printing the names and addresses of six "firms and individuals operating in Chicago as medical specialists," and it recommended "further investigation by the police department." The specialty of these doctors, as they all billed themselves, was the cure of "men's disease"—a common euphemism for syphilis and gonorrhea. All of them, the *Tribune* found, were "organized, backed by ample funds and buttressed behind expensive legal talent." It also uncovered "much evidence to indicate that their claims of curing diseases are founded largely on buncombe and that their profession is, in fact, that genial age-old profession of getting money by false pretenses."[39]

The story unfolded over the next few months, with detailed accounts of reporters' visits to "doctors'" offices. Typical of their encounters was the one with "Professor Ehrlich," whose practice was exposed in a series of articles that began in late October.

> From a quick glance at the advertisement of the 606 [the popular name for Salvarsan] Medical Laboratory at 303-4 145 North Clark Street, the average man would be apt to think that Prof. Ehrlich was in charge. He is the only man in the ad., though the picture of a Van Dyked physician adorns it. I went there today and asked for Prof. Ehrlich, and apparently couldn't understand that he was not the doctor in charge. The girl explained in a low voice, and smiling all the time, that Ehrlich was in Germany and had never been connected with the firm. She thought I was an awful boob when I asked her when she expected him back. She said, "the doctor" would be back shortly. He arrived in a few minutes—a little after 2 o'clock. Soon afterward he called me in.
>
> "Well, sir, what can I do for you?" was his greeting. "Take a chair. Sit down. Don't be nervous."

He was a rather slightly built man, but the thing that held my attention was his sunbeam hair. He wore much hair—on top of his head, down the side of his face, on his upper lip, and on his chin. He is the Van Dyke whose picture appears in the ads.

"Doctor, I'm going to be married in a few months," I said, and ran my hands through my hair quickly several times.

"I want to find out if I'm all right."

"Ever had any disease?"

"No, sir."

"What symptoms have you?"

"Why, it seems, doctor, that my hair is coming out," and I again made some rapid passes through my hair. "Some one told me that was a sign."

He laughed at this, as I imagine a man would laugh when he hears something extremely funny.

"Yes, that's a bad symptom," he declared. "It shows conclusively that your blood is impure."

"How about your throat?" which I was handling very gently. "Ever had a sore throat?"

"Yes, sir, frequently," which is true.

"Um, hum! Now, how do you sleep and eat? Regularly?"

"I haven't noticed any peculiarity, doctor."

Misleading, if not entirely deceptive, information is discovered at the outset, in the ambiguous identity of the doctor running the office. Although the woman at the front desk clearly states that Ehrlich is not there, she never gives any clear identification of the man in charge. The man, identified as "Dr. Code," appears to be a bit vague in his medical knowledge, often taking his cues from his anxious, potential patient. The reporter conveys the impression that any symptom he might identify would be a tell-tale sign of syphilis.

The interview continues:

"Oh, I can see you have syphilis," he asserted with positiveness. "I am absolutely certain of it from what you have told me. But of course it will be necessary to give you a Wassermann test for your blood. This is necessary to find out whether you have it or not and how bad it is."

"H-how much would that cost, doctor?" I asked.

"That would cost $20."

"Well, I guess I can get that much easy enough, doctor. You see, doctor, I've had a lot of debts and I'm just getting on my feet. Yes, I can get $20 easy enough."

"Where do you live?"

"In Wilmette."

"Now, it will be better for you to take the salvarsan treatment first, you see"—and he went into a lengthy discussion of my disease, and just how it should be treated. All this without laying a finger on me.

"How much would the salvarsan cost?"

"I can give you that for $30, and you may be sure that it is the genuine salvarsan."

"Thirty dollars," I gasped. "Gee, I don't know whether I can get that much or not."

Every one of the doctor's beautiful little brown hairs lay down slowly, it seemed, until I happened to think of a man I could borrow $30 from, and then he was his old genial self once more. He was sure I could get it if I told the man how urgent it was, and did I have something, a small sum, to deposit? Yes, I could afford to pay him $2.[40]

The doctor was following the current protocol for syphilis, but even if we assume that he did perform a Wassermann and would administer Salvarsan, the fact remains that although this patient has no infection, the doctor would have him receive at least one treatment. Although the doctor charges a large fee for 1913, he is suspiciously willing to settle for any amount of deposit.

After the articles about each practitioner had run for a day or two, the *Tribune* began to run letters from readers who were former patients of the men. About Dr. Code, for example, a man calling himself Bronson recounted being prompted to call at Code's office through "their advertising in the Chicago *Examiner.* The doctor told me I had blood poison [syphilis] bad. I told him I was a hard-working boy, so he arranged his payments on installments. He gave me his medicine without any directions or labels on the bottles." Bronson's suspicions aroused, he checked the credentials Code had given him, which were false, and went to another physician for a Wassermann test, which was negative. "I intended to bring a suit against them, to recover some of the money," Bronson concluded his letter, "but I didn't on account of my position and the publicity."[41] Whether "Bronson" 's fear related to being publicly named for his naiveté or for his suspicion that he might have syphilis is unknown, but either threat was sufficient for most of the patients of this and other "men's specialists" to allow these practices to flourish. For those who never suspected the deception, fear of syphilis was clearly enough to keep the doctors in business. The *Tribune* recounted stories as well of people who actually did have syphilis, who were incorrectly or inadequately treated and consequently lost large sums of money—and in some instances their health or lives.

In the course of running the series of articles, Keeley refused to accept any advertising from these and similar practitioners, whose notices had been a staple of Chicago newspaper advertising for decades.[42] It was a sacrifice to a higher standard, Keeley let his public know, costing the *Tribune* "actual and potential" annual revenue of $200,000.[43] He did not, however, point out the compensations of increased circulation, public attention, and moral prestige that followed this effort. Moreover, in terms of "higher standards," the *Tribune*'s may have been no higher than many of the other Chicago newspapers at that time. By 1912 the paper contained a combination of "sound political and financial" coverage plus "lurid crime and divorce" stories to satisfy the many interests of its readers.[44] With this series of reports, however, the *Tribune* established itself as an advocate for the public health of Chicago. It is also significant, given both the Health Department's and the *Tribune*'s assertions in 1939 of a *new* amorality in dealing with syphilis, that the series from 1913 reported on countless ordinary men who had or feared that they had a venereal disease without offering any commentary on or allusion to their past behaviors. Infected and uninfected men alike were treated as sympathetic victims, the infected men even more so, perhaps, because they were often left with a more protracted treatment or were even beyond medical help.

The *Tribune*'s most extensive moral condemnation during that period, however, was directed against prostitution. Keeley's first attack on "vice," in 1911, paralleled a similar attack from the city. *Tribune* reporters offered lively, lurid accounts of the Levee district on the South Side, where "a man cannot walk a few yards without being accosted by a dissolute woman."[45] Public outcry pressured Mayor Carter Harrison, Jr., to order the closing of the most notorious of the "resorts," the Everleigh House.[46] Vice was not so easily stamped out, however, and in the summer of 1914 several newspapers, preachers, and civic groups (including the Committee of Fifteen, which would serve as Chicago's most dominant antivice group for several decades) pressured the mayor once again to make a show of force, this time much more extensive and concerted. The Levee district was invaded once again, this time complete with a shootout between police and gangland members. The Levee was permanently closed, and Chicago's prostitution district was gone. *Tribune* reporters were with the police on the night of the shootout, July 16, and, in a battle over headlines with the other Chicago papers, the *Tribune* claimed credit for the victory over vice.[47]

"Eight to Five Syphilis'll Win"

In the first months after the enactment of the Saltiel Law, Ben Reitman's attitude toward it vacillated. His first response was skeptical. He observed that 10 percent of the people treated in the Municipal Social Hygiene Clinic stated that they had contracted syphilis from or had contact, while themselves infected, with married men or women. Obviously, he pointed out, marriage does not ensure constancy; therefore, "examining men and women before they take out a license is not enough."[48] As one more step toward universal testing, however, Reitman hailed the law, identifying it in one of his reports as one of the "More Important Factors in the Control and Decline of Syphilis."[49] In November 1937 and January 1938, he optimistically predicted, "With the damming up of Indiana and Wisconsin, easy license bureaus, the marriage rates will run about normal in 1938. . . . Illinois has occasion to be very proud of the Saltiel law. Hundreds of women and babies will be saved the ignominy and tragedy of syphilis."[50] On the other hand, Reitman feared that the Saltiel Law deterred too many people. "WHY?" he asked—because, "WE have frightened them."[51] Reitman repeated his observation a month later in a section of his report entitled "The Saltiel Marriage Law Continues to Be a Terror." He charged that "decent people" would be forced to masturbate mutually rather than marry because "there is a good deal of education yet to be done to convince our young people that it is wise to be examined before they marry."[52]

Reitman accused the Saltiel Law of fostering "syphilophobia," the fear of syphilis or any venereal disease, an educational tactic that he criticized as counterproductive.[53] Fear, however, was not without its proponents, most notably Surgeon General Thomas Parran, who once commented that "the only good patient is the frightened patient."[54] In early June 1938, nearly a year after the Saltiel Law went into effect, Reitman observed that 15 percent of the people who applied for a marriage license and then went for their blood tests never returned. In pondering why this might be happening, he advised that a physician or clerk be assigned to the Marriage Bureau to explain the nature of both the test and venereal disease to applicants, because if fear of venereal disease underlay their failure to return, "Something should be done about it."[55] "It only remains to figure out how much of a deterrent it is," he concluded a month later.[56] In the end, though, he acknowledged the success of the Saltiel Law. When the number of marriages in Illinois finally returned to pre-Saltiel levels fourteen months after passage, Reitman wrote that the law "continues to justify itself."[57]

The unexpected furor following the premarital laws calmed, to great extent, with the simple passage of time. It was also, doubtless, aided by changes in other states' laws. When the courts granted the first annulment of an Indiana marriage since the enactment of the Saltiel law, six months after the first such attempt by Maria and Edward Steinke, the Indiana bureaus had effectively been closed to citizens of Illinois.[58] Beginning with an injunction against the county clerk's office in early November 1937, and surrounded by growing pressure from the clergy, lawmakers, and citizens of Indiana, the Indiana supreme court in January 1938 finally upheld an Indiana law from 1852 that forbade the performance of marriage ceremonies for women who were not residents of that county.[59]

The news made the front-page, banner headline of the *Tribune:* "Kill Indiana Marriage Mills." The next day an editorial stated, "We are condent Indiana will soon have its own law to safeguard its own people against the venereal plague, but meanwhile Illinois, which is leading in that war, was finding its efforts being defeated so far as several thousand of its own citizens are concerned. We are very grateful for the decision." On the next page, the *Tribune* pronounced that the Illinois law was vindicated: "Marriage Tests Bring Quick Drop in Divorce Rate."[60] Six-month statistics recorded that one in twelve thousand Chicago marriages had ended in divorce, compared to one in seven hundred marriages in the six months preceding the passage of the law. Of those divorces since the Saltiel Law, twenty-one Indiana marriages were dissolved for every Chicago marriage that ended in divorce.

By the end of 1938, twenty-six states had passed laws that prohibited the marriage of infected people, although some states continued to require a test of only men.[61] In October of that year, the *Tribune* announced that eight states now had examination laws, with some states adding special provisions to prevent evasion of the law by marrying outside the state.[62] Although marriage rates remained lower than usual in Chicago for several more months, coverage of this unexpected reaction to Chicago's campaign against syphilis had more or less died out by the middle of 1938.[63]

Could anyone have predicted—or prevented—the reactions to the passage of the Saltiel and Graham bills? Clearly, the Chicago Syphilis Control Program was caught off-guard. Although enacting legislation involves a different body of people, there is some suggestion that the Board of Health might have done a better job of educating Chicagoans to the meaning and mechanisms of the laws. Could the *Tribune* have reported events differently? Certainly, a newspaper cannot think for its public; nor can it *not* cover major local events or alarms. The stories about the crowds in the Chicago Marriage Bureau at the end of June and in Indiana marriage bureaus after June, however, presented an image that contrasted sharply with the *Tribune*'s professed sympathetic stance toward men and women infected with syphilis. Their stories and editorials repeatedly implied the morality of the Illinois law by reporting on the huge profits of the "Gretna Green's" (the town's name came to be used generically for all gin-marriage mills) officials and the frivolous, often inebriated couples who patronized those places, picturing all parties as flaunting the sanctity of marriage. One upshot of this portrayal, however, was guilt by association for syphilis. Couples who sought to evade a "sober" marriage were at the same time avoiding a test for venereal disease. The sexual irresponsibility attributed to most gin marriages could not help being linked with the possibility of a sexually transmitted infection that accompanied the couples, untested, to Crown Point, Valparaiso, or other Gretna Greens.

But it was a great story: crowds, outrageous behaviors, hundreds of personal anecdotes. What newspaper could resist? The tensions between telling the news and selling the news, moreover, combine with the personal politics and beliefs of a newspaper's editors. During the first half of the twentieth century, the *Tribune*'s personality came increasingly to reflect "the liberal posture of many sons of the newly rich and powerful who had the advantages of education in the country's leading seaboard colleges."[64] The limits to which this "newly rich" liberalism could go would be seen during the New Deal era, but in the first decades of the twentieth century this personality was developing a sense of establishment, propriety, and proprietorship, thus the seeming incongruity of a growingly conservative newspaper staking a historical claim to representing enlightenment where syphilis was concerned.

Such presumptuousness made the *Tribune* a special target for other newspapers. "Long before it achieved its present 36-story Gothic pile on Michigan Av.," John J. McPhaul remarks, "the paper exuded an awesome self-confidence. . . . It has been remarked that the *Tribune* conducts itself as though it had won and not coined the slogan on its masthead, *World's Greatest Newspaper* (copyright 1908)."[65] The point here is not to attempt to assess the fairness of the accusations and criticisms leveled against the *Tribune*. The paper's evolution in the journalistically freewheeling city of Chicago, however, as well as its history of reporting on venereal disease and its policy about serving its community (albeit according to the newspaper's own moral predilections), contributed to the role that the *Tribune* would play in the Chicago Syphilis Control Program.

Jokes about the *Tribune*, however, reveal real, ideological differences among its owners and editors, the designers and administrators of the Chicago Syphilis Control Program, and many citizens of Chicago. For example, Lawrence Linck told an anecdote of a man who came into one of the city's blood-testing stations that humorously symbolized the *Tribune*'s reputation among many of the working-class Chicagoans—the very people for whom the free testing clinics were chiefly created. "I want syphilis," Linck quoted the man as demanding. The doctor corrected the man, telling him he wanted the syphilis *test*, but the man persisted in his original demand. The doctor tried again,

> "My man, don't you know that syphilis is a disease, a bad disease?"
> "Just give me syphilis; can't you do it?"
> "Yes, it is possible to do it, but—"
> "Well, since the Tribune is against it, I am for it."[66]

A second, similar, joke appears as a footnote in McPhaul's *Deadlines and Monkeyshines* to illustrate his observation that the *Tribune* perennially backed candidates and issues that were voted down or out: "Historically the *Tribune* has been read but not followed. Its candidates and pet issues rarely win at the polls. Hence the jest: 'The *Tribune* has come out against syphilis. Bet you 8 to 5 syphilis'll win.' "[67]

Officials with the Public Health Service noted this attitude toward the *Tribune*, too, and analyzed it in their internal correspondence. Reporting on a visit to Chicago, where he attended, among other things, the *Tribune*'s much-publicized testing of their own staff, the Public Health Service's special consultant for syphilis in industry, Albert E. Russell, wrote to Parran,

> The Chicago Tribune was having the bloods taken from their employees while I was in Chicago. This paper has done a splendid bit of educational work on syphilis. . . . It is interesting that the Tribune has been the only Chicago paper to do this. There are many people who do not like the Tribune for various reasons and particularly those who are supporters of the present Federal administration, in as much as the

> Tribune has not let an opportunity pass to criticize the President and his policies. It seems unfortunate that the other papers in this big city have not done anything about syphilis publicity.[68]

The *Tribune*'s mixed loyalties would become more problematic to the newspaper itself in later months. Although the tension between the paper's readership and the goals of the Chicago Syphilis Control Program were evident to the program's leaders from the early months of its operation, any perceived conflicts of interest on the part of the *Tribune* did not hamper the newspaper's initial conviction that it had an important part to play in the program's success.

CHAPTER FOUR

Taking the Rap

Early in the work of the Chicago Syphilis Control Program, a story in the *Chicago Tribune* reported that women between the ages of fifteen and thirty-four comprised the largest group of people responding favorably to the syphilis questionnaire.[1] Further information about these women, such as marital, socioeconomic, or work status, was not known, but the *Tribune* interpreted this response as coming from the "girl next door," whose innocence—and good hygiene—were above suspicion. This group's response was not only offered as additional proof that the taboo surrounding syphilis had been lifted but also praised as the epitome of high-minded civic duty, as America's daughters put the health of the nation above false modesty.

One might question the efficacy of an undertaking whose most ready participants were perceived to be at particularly low risk for having syphilis; on the contrary, these women's enthusiasm became the foil for another story. At the close of this article, reporter Arthur Evans briefly mentioned that Judge E. J. Holland, who presided over the Woman's Court, had sent to the Chicago House of Correction four women with active venereal infections "and asked that they be isolated for treatment during the period of their sentences." Formerly called Morals Court (an interesting word change in itself), Woman's Court dealt mainly with women arrested for prostitution and the men found with them at the time of arrest. The article's juxtaposition of large numbers of Chicago's "good" young volunteers and the prostitutes tested at court order reveals a coercive flip side of the banner-waving of public cooperation and introduces the third, and briefest, series of stories that ran in the *Tribune* during the summer of 1937: the legal actions that accompanied the antisyphilis campaign.

One goal of the Chicago Syphilis Control Program was to test all of the city's citizens for syphilis. Although the citywide testing program depended heavily on voluntary participation, it also counted on legal support of its goals. It applauded the Saltiel Law as an assurance that one portion of the public would be guaranteed a test. It supported the passage of a similar law to mandate prenatal testing of pregnant women. It also encouraged policies or regulations in hospitals, industry, and higher education that would mandate examination of all patients, employees, or students. Finally, it resurrected old city ordinances that permitted court-ordered testing of selected individuals.

In a drive to discover and treat all cases of syphilis, both mandatory and voluntary testing present political and practical problems. Besides issues of individual rights, the prospect of compulsory universal testing raises questions regarding logistic and economic feasibility. A purely voluntary program, on the other hand, offers no guarantee that its goals will be met, even in the short run. A voluntary program with mandatory testing of targeted groups, as described in this and the preceding chapter, however, differentiates among various groups of people. In the case of testing for syphilis, that differentiation undermined the neutrality toward the disease—and the diseased—that the Chicago Syphilis Control Program professed to exercise.

"If They Do Not Look Responsible"

The article about Chicago's civic-minded younger women appeared as almost a final footnote to a drama that had been playing itself out in the pages of the *Tribune* for the preceding month. In late June, the paper reviewed all the Chicago and Illinois statutes regarding venereal disease, noting that in 1919 judges were empowered to order medical examinations "for persons suspected of having venereal disease and authorizing health authorities to segregate them for treatment."[2] Stanley Armstrong did not tell his readers, though, that this law had passed into relative disuse following the antisyphilis drive of 1922 through 1926, largely through the efforts of reform groups that opposed this punitive approach to "redeeming" prostitutes.[3]

Whatever the history of the law's disuse, in the summer of 1937, three judges in Woman's Court made a plea through the *Tribune* for legal authority to order medical treatment for women brought before them on charges of prostitution, an authority that in actuality they already possessed. Judges Joseph B. Hermes, Matthew D. Hartigan, and Gibson G. Gorman reported that in the past few days twenty-two prostitutes known to have syphilis had left the jails after their fines had been paid by their pimps or boyfriends. Although the women had been examined upon their arraignment, they had only been fined and urged by the judges to seek treatment at Bridewell, the county jail.[4]

The judges' plea was followed two days later by a *Tribune* editorial entitled "Carriers of Venereal Disease," which called for the Chicago Board of Health to apply its quarantine laws to venereal disease.[5] It was two weeks before the *Tribune* reported again, this time from Arthur Evans, that a state statute of 1919 allowed judges to "require the examination and treatment" for venereal disease of *anyone* brought before their court whom the judges had "reasonable grounds" to suspect to be infected with a venereal disease. Evans also repeated that judges could "segregate" any person discovered to be so infected.[6] Two weeks later, the *Tribune* noted the forced hospitalization of prostitutes.

The revival of the old law, though, took a new turn. The resumption of this exercise of the judge's prerogative soon spread to other venues besides the Woman's Court—a move believed by both judges and reporters to be unprecedented throughout the United States. Judge Oscar S. Caplan of the Des Plaines Street Court, an area largely populated by transient men, stated that all men ordered to be tested would be granted a continuance of their trial for a week while awaiting the results of their test. If infected, they would "be fined an amount sufficient to keep them in the Bridewell for clinical treatment until they are cured or rendered noninfectious."

Caplan based his decision on statistics that showed an extremely high rate of syphilis in postmortem examinations of residents who died in the Des Plaines neighborhood, but his statement is puzzling.[7] Many of the transient men arrested in the Des Plaines Street area were probably older men whose transience had contributed to their lack of previous health care and whose syphilis was most likely in a noninfectious stage. Not all "infected" people are "infectious." Moreover, because "cured" and "noninfectious" are not necessarily the same thing in the case of syphilis, it is hard to determine the exact basis for compulsory hospitalization. Finally, if men were being held until cured (although, given the length of treatment required for cure at that time, this seems unlikely), many people past the infectious stages of syphilis were being isolated in the name of the public good, when the only people they were directly endangering was themselves. Although there are obviously some gaps in the medical logic of this story, its general effect on a relatively uninformed readership would probably foster the idea that all syphilis is infectious. Indeed, none of the articles appearing in the *Tribune* during this period ever discussed the distinctions between the various stages of syphilis, infectiousness and noninfectiousness, creating an impression that all syphilis was infectious, virulent, and menacing.

The next day the *Tribune* reported, "First Men Take Syphilis Tests at Court Order." "The single standard" created by this new exercise of an old law, reported Arthur Evans, "removes an objection raised from some sources that women suspected of having syphilis were being treated while men were escaping."[8] Unlike the interview with Judge Caplan, though, this article printed an interview with Judge Holland of the Woman's Court, who reported that, of the ten to fifteen men and women tested from that court daily, only those people found to have syphilis in a *communicable* stage were subject to mandatory confinement to the contagious disease hospital. A week later, Evans, returning to the Des Plaines Street Court, reported that one-fourth of all prisoners examined in that court, eighteen of seventy-six, were found to be infected with venereal disease. They were sent to Bridewell for fifty-five days, where they were "working out" their fines of $100 plus costs. Again, though, the lack of distinguishing between communicable and noncommunicable syphilis could be misleading. The Des Plaines Court judges were commended for their actions by Chief Justice John J. Sonsteby of the Municipal Court.[9]

When statistics at the end of August revealed that 10 percent of all prisoners examined in August had "a venereal disease," Chief Justice Sonsteby "announced that clinical examinations of defendants suspected of carrying venereal diseases would be enforced henceforth in all the branches of the Municipal court hearing cases involving possible suspects."[10] Although the extension of the testing statute beyond the Woman's Court was a more equal exercise of the old law, Sonsteby's decision to have the judges order tests only on those defendants "suspected of carrying venereal diseases" still exercised an arbitrary, potentially discriminatory or punitive power that contradicted the desired image of an egalitarian, nonstigmatizing disease.

Evenhanded as the law prided itself in being, it placed substantial discretion in the hands of judges. Herman Bundesen emphasized that the testing of men was new and progressive, telling a *Tribune* reporter that "Judge Holland is treating women and the men brought in with them on equal terms, and is showing what can be done in the courts without overreaching the power of authority." Holland himself, however, talked to Evans exclusively about prostitutes. "If they refuse [to be tested]," he was quoted as saying, "I raise their bonds prohibitively high. If they are found afflicted, but look responsible and agree to take medical treatment to render the disease noncommunicable, they are put on probation. If they do not look responsible, or demur, they are sent to the Bridewell with a recommendation they be hospitalized." [11] For all of Bundesen's reference to testing men, when judges in the Woman's Court were interviewed they talked almost exclusively about women. Whatever the liberalization of the intent an execution of the law, women were probably still being judged more harshly than men for their sexual actions.

"The Social Evil" in Chicago before 1937

Chicago's early efforts against prostitution are inseparable from its various campaigns against venereal disease. The first decades of the 1900s were a time of national organization against prostitution, and Chicago was one of the many cities that formed vice commissions to study the "social evil." It was, moreover, the first city to form such a commission and publish its report, *The Social Evil in Chicago,* which went through several printings and was distributed nationally by the American Vigilance Association.[12] Members of the commission, appointed by Mayor Fred A. Busse, included William A. Evans, Louis Schmidt, and numerous other physicians, judges, members of the clergy, professors, and leading philanthropists and businessmen.

The language of the report is instructive. Its aim was clearly to strike fear and revulsion in the minds and hearts of readers. It took a clear moral stance against prostitution, creating an amazing metaphor that linked pregnancy, prostitution, and disease. The opening paragraphs, which allude to but never name syphilis, set the tone:

> The honor of Chicago, the fathers and mothers of her children, the physical and moral integrity of the future generation demand that she repress public prostitution.
>
> Prostitution is pregnant with disease, a disease infecting not only the guilty, but contaminating the innocent wife and child in the home with sickening certainty almost inconceivable; a disease to be feared with as great horror as a leprous plague; a disease scattering misery broadcast, and leaving in its wake sterility, insanity, paralysis, the

blinded eyes of little babes, the twisted limbs of deformed children, degradation, physical rot and mental decay.[13]

These exclamations were written in the hey-day of "yellow journalism," when superlatives and extremes were a familiar part of the rhetorical armamentarium. Nevertheless, their language is deliberately sensational, the moral judgments clear.

The report was considered a model of epidemiological thoroughness for its time. It is 353 pages long, with appended tables and charts on all aspects of prostitution in Chicago. It describes the various facilities in which prostitution occurred around the city: houses, flats, hotels, lake boats, Turkish baths, dance halls, and tenements. The number of each of these sites throughout the various areas of the city is reported. The report looks at prostitutes' use of alcohol and drugs, the occurrence of venereal disease, other crimes associated with prostitution, and sources of "supply." One chart records the careers of thirty prostitutes, noting their current ages, ages upon "entrance to life" of prostitution, former occupations, weekly wages, "house prices," number and sexes of siblings, use of their income, and why they became prostitutes.

The vice commission made two recommendations: to appoint a morals commission and establish a morals court. Both of these bodies were created. It is significant that although the commission took an unabashedly moral stance, it viewed its recommendations as the result of an evenhanded, open-minded process. It reported that "the Commission has squarely faced the problem. It has tried to do its duty by placing before the public the true situation in Chicago. It presents recommendations carefully and conscientiously drawn." It went on, pointedly, to describe the commission's "attitude." It attested, "At all times, while honest in the statement of conditions, it has assumed an ultra-conservative attitude in its criticisms. It believes that only through such an honest and conservative study can the true situation be given to the citizens of the city."[14] The vice commission's assertion of fairness and methodical nature is similar to the enlightened approach to syphilis proclaimed by Thomas Parran more than twenty years later. Such parallels may not so much demonstrate historical or moralistic blindness on the part of its speakers as they do the changing methodologies of science or social science that later reveal the flaws of past methods, even while being held imaginatively captive by present ones.

The eventual closing of the Levee district, a result of the work of this commission, has already been recounted, as has the next highly visible city action against prostitution, Herman Bundesen's antisyphilis campaign of the 1920s. Bundesen began the campaign as a cooperative assault, with his initial attempt to work with houses of prostitution. "If you won't be moral you must be clean," he chided, and put force behind his words by raiding the houses, testing the women who were arrested, and forcibly hospitalizing infected women who did not voluntarily put themselves under treatment.[15] On occasion, city officials quarantined houses where infected prostitutes were known to work. "Yet," reported the Division of Social Hygiene, "in spite of the determined stand on the part of the administration to wipe out prostitution, in 1926 there were 5,213 women arrested charged with prostitution. Each one of these was examined by the Health Department."[16]

Fourteen percent, or 761, of the women arrested in 1926 were hospitalized. Women who were diagnosed with a chronic gonorrheal condition, often with a tubal or ovarian complication, or women with positive Wassermanns but no evidence of epidermal rashes or throat or vaginal lesions were sent to clinics or private physicians for treatment. "The present policy of the Health Department is to isolate only those cases where the preponderance of laboratory and clinical evidence indicates that they are infectious," read the annual report of the Division of Social Hygiene. This statement seems to be contradicted a few lines later by "the Health Department has adopted the policy of compulsorily hospitalizing all prostitutes with *suspicious* lesions, realizing that one syphilitic prostitute is capable of infecting a hundred men in a very short period of time."[17]

It was during this period that Ben Reitman became heavily involved in working with prostitutes. Some of his writing about that time in Chicago's history provides additional information and insight into the lives of prostitutes during the 1920s and 1930s. Reitman described Lawndale, the infectious disease hospital, in *Sister of the Road: The Autobiography of Boxcar Bertha as told to Dr. Ben Reitman.*[18] It is a "sociological novel," fiction interspersed with statistical information about hoboes and prostitutes, names of actual Chicago officials, and accounts culled from his own experiences and those of many of his friends. Bertha (a fictitious character composed of experiences from the lives of several women—and from Reitman's life as well) is arrested for prostitution and found to be infected with both syphilis and gonorrhea. She is sent to Lawndale, about which she recounts,

> The six weeks in Lawndale sped along. The matron let me help with the clerical work so I was busy and cheerful. One would suppose that a venereal disease hospital would be a very sad place. The opposite was

> true. We danced and played dominoes and checkers. We had all kinds of books and magazines and newspapers. There was plenty to smoke and every day large bundles of food were sent in to the girls. We were not allowed to have visitors, but all our pimps drove up in their cars and stood on the side walk and waved to us. Most of all girls received and wrote letters.[19]

Bertha's description of her weeks at Lawndale are presented with greater cheer and equanimity than was recounted to Reitman by Margaret Parker (the "Grace Kelly" quoted elsewhere), but Bertha hints at a darker, personal side of a woman's stay at Lawndale when she adds, "All of the girls wanted to get out of Lawndale. All were worried about what their pimps were doing in their absence. All showed they were afraid of losing their men."[20]

The tenuous security of a prostitute's existence, dependent on domineering men and a judgmental social system, leads the outspoken, flamboyant Bertha to abandon prostitution for less repressive work. But other women, Reitman was aware, did not always possess such freedom of choice or movement. He depicted the special situation of women with venereal infections in the *Outcast Narratives*, a manuscript of more than a hundred unpublished poems, most of them about his friends and patients. The structure of the *Narratives*, with the creation of a persona for each poem, clearly echoes the plan and style of Edgar Lee Masters's *Spoon River Anthology* of 1915, but the short line and terse, often deliberately flat expression are Reitman's own. "Charlotte" is typical of the style and tone of the poems, with the closing line in the character's own words containing not only her philosophy of life but also her spirit.

> Charlotte was a cute, bob-haired girl,
> Eighteen, spent most of her life in the country.
> She came to the city and lived with her sister
> Who was fast, and her brother—a gambler.
> She began to frequent a part with a bad reputation
> And imagined she was having a good time.
> She gave Don her love and received a dollar and a half.
> Dave took her love and the money.
> She was caught with Johnny in the park
> And sent to the detention home.
> She escaped and began to ply her business
> But a sailor infected her with a disease

And she went to the hospital for two months.
After that the police watched her
So she married a night engineer,
And said, "I still go to the park for fun and expenses."[21]

Charlotte's story is a familiar one among the *Narratives*—the young innocent from the country swept away by the color and excitement of the city. Charlotte, however, is resourceful, practical, and independent. She shows no remorse for her life, but rather an honest openness about her activities and the pleasure and profit that prostitution continues to provide her. Reitman presented the effects of disease on Charlotte's life and also pointed out the connection between Charlotte's hospitalization and subsequent police harassment, implying that their pressures to control her conduct were both intrusive and ineffectual.

The dangers to a prostitute's—or *any* sexually active woman's—physical health were only one of the threats that venereal disease posed, as Reitman spelled out in his poem about "Evelyn":

Evelyn was a working girl.
Twenty-four, pale and spiritless
She lived at home with her parents.
She had a sweetheart named Joe
And every Saturday night they would "go to a dance."
When the police raided the Revere House
They got twenty-two couples besides Joe and Evelyn.
The Health Department examined Evelyn
and sent her to the hospital for five weeks.
They let Joe out on bail the same day.
Evelyn's parents never knew what became of her.
When their case came up in the morals court
The Judge fined Evelyn fifty dollars.
She didn't have the money, so she went to jail.
Joe was fined five dollars which the judge remitted.
When Evelyn finally got out of jail
She said: "Men always give women the worst of it."[22]

Evelyn, it should be noted, was not a prostitute, but houses of prostitution at the time often rented rooms to couples. If the houses were raided, the women in these couples were arrested and treated the same as the prostitutes. In this poem, Reitman points out the inequity in punishing only half of an infected population. Evelyn's words upon her release from jail recognize the injustice, but they are deliberately ambiguous. She implicates not only Joe in her charge against men but also the doctors and judge, whose system punished only her.

"All a Part of Their Day's Work"

A few days before the *Tribune* reported the three judges' plea for the authority to hold infected prostitutes, Reitman urged the Woman's Court to be more rigorous in ordering infected women to be hospitalized, but he also advised "the Judge to fine or sentence a number of men who are caught in whorehouses attempting to buy sex while they have an infectious venereal disease. If this were done, and publicity was gotten for it, it would help considerably."[23] When the first men were fined and ordered to be held, though, they came from Des Plaines Street rather than Woman's Court. These men were generally not engaging in illicit sex but were drunk and obstreperous—and with no place to go until their drunkenness and obstreperousness wore off.

Although Reitman applauded all steps in the direction of universal testing, his enthusiasm for this renewed scrutiny was tempered at times by skepticism about its efficacy. "Not for years have we seen such a large percent of the women from the Woman's Court examined," he commented, for example, then added that he had heard that some of the women being jailed because of infections had not been receiving treatment. He offered to look into those charges himself. Reitman initially expressed hope that this new and renewed vigilance would serve syphilis control well, but he was still skeptical that the judges' actions, sensational as they might be, could actually stop the spread of syphilis through prostitution—the same conclusion that the Division of Social Hygiene had reached in 1926.

In June of 1937, Reitman crowed that a police raid of the Best Hotel—the first such raid of this well-known house of prostitution in two years—had led pimps to say, "It's too hot for us." Reitman praised the action, commenting that the "pimps take the raid of these houses quite seriously."[24] Less than three months later, though, he changed his opinion. Houses had quickly accommodated the new regime. "Talking with a group of discouraged pimps, ropers and runners, and cab men who specialize in line loads," Reitman learned that recent pressure from the Committee of Fifteen and other reform groups had not hampered business. "Little Joe," who "specialized in distributing whore house cards to taxi cab drivers," scoffed at that notion, and told Reitman, "Nothing's closed up. Every joint that was closed is open. If they don't open in the same spot, they start business next door or across the street. The town is wide open."[25] Most of the houses owned by the syndicate were still running because they simply closed during the hours that the police usually performed raids. Besides accommodating the raids, Reitman also pointed out, managers of houses knew that police would never pursue the raids ruthlessly because the houses were good sources of information; in order to stay in business, large houses frequently functioned as willing "stool pigeons."[26]

Moreover, mandatory hospitalization of prostitutes was not a strong deterrent to prostitution. Reitman concluded, "The managers and owners of joints don't care how many girls or customers are sent to the House of Correction, because they can get new ones, but if their joints would be placarded with a sign, 'Venereal Disease. Keep out,' they would be interested. Raids and pinches and fights and syphilis are all a part of their day's work."[27] A month later, in October 1937, the *Tribune* reported that the west-side flat of Mrs. Ruth Dorker at 401 S. Paulina had been served with a quarantine. Dr. Bundesen, hammer in hand and accompanied by Dr. George Taylor, was pictured in front of a door, affixing a square sign with three lines of large print:

GONORRHEA
Venereal Disease
KEEP OUT

"Quarantines will be used only against persons who refuse to aid in preventing the spread of venereal disease," he said, and Dr. Louis Schmidt was reported to have called the procedure "the usual one in preventing the spread of other infectious diseases."[28]

Such actions during the highly publicized measures of 1926 had resulted in sharp, widespread reprimand of Bundesen. A similar reaction may have followed his act in 1937, because no subsequent reports of quarantine followed this one in the *Tribune*. The first annual report of the program, however, suggests that such activities did not cease altogether: "All delinquent patients are visited by a 'follow-up' investigator. The patient is warned that he or she has violated the Board of Health Regulations and that a second offense will mean quarantine. A patient who breaks such quarantine is subject to arrest and sentence to the House of Correction. Placarding and arrests have resulted in several instances and this method seems to be more effective."[29] In its annual report for 1939–40, the Syphilis Control Program cautiously noted that "quarantine measures" for houses of prostitution were "utilized occasionally," with the "temporarily good effect in returning delinquent patients in the neighborhood to treatment."[30]

A year and three months later, Reitman wrote his only report of quarantining a house himself. He wrote of placarding a house on Milwaukee Avenue after an investigator was told by the "madam [that] she knew nothing about" the infected prostitute who reportedly worked there. The sign was torn down and replaced, and the madam threatened Reitman by telephone and letter, "but," he wrote, "the next day the woman who was the source of infection reported to the clinic for treatment."[31] The upshot of Reitman's actions is not easily determined. His subsequent reports and correspondence from the period indicate no overt censure or punishment for his action, indicating that either the practice was common enough not to excite undue reaction or that he had followed correct procedures to assure that his act was merited.

Reitman's desire to isolate prostitutes infected with syphilis may seem incongruous with his professed sensitivity to their inequitable treatment in the legal system—but not so incongruous with his more negative statements about prostitutes in general. Moreover, Reitman was a firm believer in the tenets of public health, for which quarantine was a common—and widely accepted—practice at the time. He took quarantine one step further when he recommended the mandatory testing of all women who worked in, or even frequented, taverns; he was quick to add, however, that not all women who did so were prostitutes. "The tavern," as Reitman described it, "has now become a social institution, a neighborhood recreation center, a popular place of amusement, and we do not wish at all to be an alarmist, but . . . the larger the consumption of alcohol, the more venereal disease."[32] More than once he recommended placarding taverns where employees or regular customers—identifying only women, however—were known to have infectious syphilis.[33]

While Reitman's statements do reveal a bias, his recommendations were also based on his acknowledgment that women, as well as men, were sexually active outside marriage, whether for money or not, whether working for a pimp or not. His comment about the inevitability of syphilis in prostitution was made with a straightforwardness that approaches syphilis, prostitution, or any sexual exchange in a businesslike way—syphilis as an occupational hazard, or, to push the metaphor even further, syphilis testing as quality control. In that light, and alongside the standard use of quarantine in public health at the time, such actions would be logical and reasonable. Such an attitude, however, requires an acceptance of prostitution and sexuality that most campaigners against syphilis were not willing to concede.

"Not a Badge of Disgrace or Immorality"

In the logic outlined above, *spreading* disease is immoral but *having* disease in and of itself is not. Such a differentiation in the case of syphilis, however, rests upon an underlying acceptance of the irrepressible inevitability of sexual expression itself, whether licit or illicit. Reitman believed in the naturalness and essential blamelessness of the "sex impulse"—in fact, he celebrated it. On the other hand, Surgeon General Parran's attitude about syphilis and people who had a venereal disease was paradoxical. He made statements such as the following: "It cannot be repeated too often that first and foremost among American handicaps to syphilis control is the widespread belief, from which we are only partially emerging, that nice people don't talk about syphilis, nice people don't have syphilis, and nice people shouldn't do anything about those who do have syphilis."[34] Yet, as he continued to argue for open-mindedness, a qualification slipped in. "Before we are capable of teamwork," Parran continued, "all of us together—physician, public official, citizen—must learn to think of syphilis scientifically as a dangerous communicable disease, which it is; rather than moralistically as a punishment for sin, which it *often* is not."[35] Thus, Parran said, syphilis *is*, sometimes, a punishment for sin.

Both Parran's understanding of syphilis and his proposal for controlling it reflected his moral judgment. He acknowledged four ways to prevent infection: "Moral prophylaxis, which involves the social regeneration of a whole people, mechanical protection, emergency chemical disinfection, or the medical treatment which sterilizes open lesions."[36] He conceded that all four methods have their uses, but he considered chemical prophylaxis to be an "emergency" measure. Although Parran admitted that achieving moral prophylaxis for venereal disease would require "the social regeneration of a whole people," he saw it as the ultimate goal not only for the eradication of syphilis but also for the creation of a better world:

> For it must be made clear that moral prophylaxis can not be neglected merely because the sum total of its ideals is not attainable here and now. It is of the utmost importance for us in the United States to

> encourage the education of our young people to decent living through all the means at our command of church, school, official, and voluntary agency. . . .
>
> Moreover, I do not believe that sexual morality should stand or fall on its relationship to venereal disease. I hope that my sons will practice it because they understand and want the advantages of the better way of life; not because they are afraid of syphilis if they fail to practice it.[37]

Having syphilis was "not a badge of disgrace or immorality," the *Chicago Tribune* proclaimed on June 14, 1937, as it set out to enlist the support of all its citizens in fighting syphilis.[38] Even with the best of intentions, however, the *Tribune* did not always send a clear message about syphilis. The paper, however, was not the only party caught on the horns of a metaphorical or moralistic dilemma. Herman Bundesen himself moved quickly from tolerance to accusation when he observed, "Nice people do get syphilis. And I say the difference between those who do and those who don't is misfortune, and nothing else. The syphilis carrier is a potential murderer, and I say he must be stopped whether he likes it or not."[39] Although his distinction between a "misfortun[ate]" person and a "potential murder" probably turned on a conscious decision to be tested or to comply with the prescribed treatment, his unqualified statement tips the scale toward blame more than forgiveness. Bundesen's words, moreover, were accompanied by a chart showing the spread of infection from one fifteen-year-old girl, with no concern expressed for how she came to be infected in the first place.

A perfect example of the paradoxes that can be created in reporting the news occurred in the very design of one of the pages of the *Tribune*. On June 18, 1937, the major article on that page carried the headline: "Tribune Pioneer in Open Fight on Syphilis Plague; Frankness First to Break Venereal Taboo." The headlines of a shorter piece on the same page read, "Syphilis' Effect Is Key in Murder Trial of Woman; Husband Unbalanced, Her Defense Contends."[40] The second article, from Manitowoc, Wisconsin, told the story of a woman accused of killing her husband in a fight that ensued after she discovered him in bed with another woman. The husband, the story went on, became enraged when his wife told the other woman that he had an infectious case of syphilis, and he attacked his wife. Although the woman's lawyer defended her on the grounds that his infection had affected his brain and thus put her in mortal danger, four previous tests for syphilis had been indecisive, two of them being positive and two negative. It was unclear, moreover, whether his syphilis, if he even had it, was infectious, noninfectious, or completely cured. As the story continued, it was revealed that his rage was not related to any form of neurosyphilis, as the headlines might be read to intimate, and his "unbalanced" state was triggered by his wife's interference rather than the effects of the disease. The headline is misleading in this instance; syphilis is connected to marital infidelity—although the source of the man's possible infection is never revealed. However openly syphilis may have been discussed, its connection with immoral behavior was surely made by many readers.

Two other articles further illustrate this association. They also reveal that the discretion permitted judges in ordering a test for syphilis was often exercised more for punishment than precaution. The first reported instance of a judge requiring venereal disease testing under the new vigilance of the Syphilis Control Program actually preceded the testing of transients in the Des Plaines Street court. At the end of a news article about the overwhelmingly favorable response the syphilis questionnaire was receiving, the following, unrelated sentence appeared: "Meanwhile, it was learned that suspects who were held over night at the county jail after the recent roundup of suspected medical quacks were given Wassermann and other tests for venereal diseases."[41] Contrary to the testing of transient men, whom studies suggested to have high rates of syphilis, or men and women brought before Woman's Court where testing for syphilis had historical precedence, no reason was offered here to justify testing the "quacks." If these men were claiming to cure syphilis, as were the men who were the subject of *Tribune* exposure in 1913, then the story might signify that association with syphilis alone was sufficient to warrant testing, but the *Tribune* did not specify the claims of these "quacks." Ordering the men to be tested seems more intended to impugn their claims to medical expertise, but a secondary consequence could also link syphilis and allegedly criminal behavior.

The second story appeared a month later, in early September, in a brief flurry of scandal. Thirty-year-old Philip Robinson Yarrow, "son of the professional vice crusader, the Rev. Philip Yarrow," was arrested on charges of disorderly conduct when an irate husband returned home to find him in bed with his wife. The newspapers reported that both Yarrow and Mrs. June Stadelmann were ordered to be tested for syphilis and gonorrhea after being brought up before Woman's Court. Mr. Stadelmann was said to have "indicated that he would dismiss the charge against Yarrow at the time if he is found to be free of venereal disease."[42] His reason for setting this condition was not given. Was Stadelmann's concern for his wife's health or for his own? Is having venereal disease a greater crime than marital infidelity?

Moreover, would the story have ever made the newspapers had young Yarrow's father not been a well-known leader in antiprostitution work? No evidence was given—nor was the suggestion even made—that Mrs. Stadelman was having sex with Yarrow for money. Whatever the reason, ordering Wassermann tests for Yarrow and Mrs. Stadelmann, at a time when most couples being tested were involved in prostitution, seems a conscious rebuke or taunt to a young man who, the judges were implying, should have paid more attention to his father's preachings. And, again, it further implies that syphilis might logically accompany—or justly punish—such behavior.

In both of these cases, judges' selectivity in whom to order to be tested carried an accusation of guilt, which was directly tied to the possibility of having syphilis. Ben Reitman argued that the only way to overcome such implications was to test even more people. The only time that Reitman was mentioned in the *Chicago Tribune* in connection with the activities of the antisyphilis campaign occurred during this period of renewed, court-ordered testing. In August 1937, when the men from the Des Plaines Street Court were tested, the *Tribune* reported that the men "were taken to the clinic by Dr. Ben L. Reitman of the Chicago Society for the Prevention of Venereal Diseases."[43] Reitman's reports provide additional information about this event. "Five prisoners were sent in the Patrol Wagon to the Municipal S.H.C. [Social Hygiene Clinic]," he wrote, "and two of the men were found to be suffering from gonorrhea." One of the men, already a patient at the clinic, was released after promising to continue receiving treatment there. The other infected man, "an old man over fifty, who was arrested for exposing himself to young girls," stayed overnight at the police station. The next day he was fined $200 and sent to the House of Correction.[44]

Reitman had apparently been attending the court regularly, because three days later he submitted the following lively report:

> Judge Caplan continued to let us sit by his side to make suggestions as to who should be sent to the Board of Health. The big colored man with an ugly looking 44, a revolver, was discharged, but we suggested that he be sent to the Clinic, because there is a distinct relationship between violence—suicide and homicide—and syphilis. The wife-beaters—men who brutally beat their frail wives; syphilis and alcohol make men irritable. They're a bad mixture. All the men who were not sent to jail were recommended to the clinic. The robust, hard-working, casual laborers who came in to spend or be robbed of

> the money they earned working on the railroad—they got drunk, were with some women that they picked up in a tavern. All they can remember is that they got drunk, had a fight, and woke up in the police station. There are many old, habitual drunkards, over 60 years old, to whom life is a thing of the past and a love life is nigh unto impossible. No suggestions is made about these.[45]

Reitman thus urged Judge Caplan to test nearly every man brought before his court, although his dismissal of the "old, habitual drunkards" might be a bit abrupt. Reitman saw this expansion as just the first step in a more sweeping, even less discriminatory policy. "If the Speeders [Traffic] Court could be induced to send part of the speeders up for examination," Reitman recommended around this same time, "that would help and would be very splendid publicity. Also, the Court of Domestic Relationship ought to send many entire families up for blood tests. We have made a beautiful start and, before the year is out, *every person arrested for any cause*, will be blood-tested, if we go about it in the right way."[46]

The "right way" for Reitman called for measures far more sweeping than those advocated by the *Tribune*'s editors or implemented by the courts. By openly expanding court-ordered testing into dockets that dealt with charges other than prostitution and vagrancy, syphilis might be seen to accompany no particular crimes and thus diminish the stigma attached to it. When mandatory testing, as was intended, spread into schools, hospitals, and industry, then testing in the courts would become part of a wider, routine screening, whenever and wherever institutionally convenient, for an infectious disease for which the lack of a vaccine made repeated testing necessary.

During the summer of 1937, the quick, easy end of syphilis was seen as a simple—albeit sizable—task. Government officials and citizens vowed to achieve this goal by several means. Although voluntarism was promoted as the most laudable means, civic and medical leaders supported compulsory ones as well. These enforceable measures, however, at times seemed to pose more of a threat to citizens than an opportunity. How was it possible to urge responsibility without implying irresponsibility, to convey urgency without creating fear or animosity, to advocate rather than accuse, to achieve compliance without coercion? Such questions would continue to challenge officials of the Chicago Syphilis Control Program as its work moved beyond its entertaining, fast-paced beginning.

CHAPTER FIVE

Dragnets

After great public fanfare, both deliberate and uncourted, the medical offices and clinics of Chicago opened their doors to an anticipated flood of volunteers for syphilis testing. Instead of opening on October 1 as planned, the Chicago Syphilis Control Program responded to the increased demand for tests that was already being felt in city clinics and laboratories and began its official work on September 1. Despite the sharp drop in the number of marriages after July 1, testing of premarital couples accounted for much of the new business, as did the resumption of mandatory testing of men and women through the Chicago court system.[1] More important, the increased number of people being tested yielded a higher percentage of infections than had been estimated.[2] In early May, Ben Reitman reported that there had been "more visits to the [Municipal Social Hygiene] clinic than any other quarter in the twenty years the Venereal Clinic has been established."[3]

The increases in both tests and infections discovered were enough to convince the program's officials that any delay would be unnecessary and inadvisable. "City-Wide Tests for Syphilis to Begin Tomorrow; Date Moved up as Result of Ballot Enthusiasm," the *Tribune* announced on August 31, adding, as evidence of the importance of the moment, the arrival of Paul de Kruif, special advisor to Mayor Kelly's Committee of Four Hundred, for a meeting with the program's key people in the writer's private apartment at the Drake Hotel.[4] The next day, Dr. Reuben Kahn arrived in town as well. The "noted bacteriologist of the University of Michigan" and inventor of the Kahn blood test, which was replacing the Wassermann as the first test for syphilis, would oversee the installation of his new test and the standardization of its use in Chicago laboratories.[5] The real work of the Chicago Syphilis Control Program was about to begin.

"*To Their Own Physician*"

On the first day of citywide testing, business was brisk. "Response Heavy as Tests Start in Syphilis War," reported the *Tribune*, noting that, "Aside from those given by private physicians, tests were given yesterday at 28 maternity centers of the city board of health, at the laboratory in the police building, 1121 South State Street, and at the board of health laboratories on the 7th floor of the City Hall."[6] The "aside from," however, was significant. From the planning stages of the Chicago Syphilis Control Program, the cooperation of physicians, many of whom belonged to the conservative Chicago Medical Society, was deemed essential. For one thing, case reporting was crucial to the success of the program, and Chicago's private physicians had never been good at reporting their patients who had syphilis and gonorrhea, city law notwithstanding. For another, although people could be tested for free in the program's own clinics, its money would go farther if as many people as possible were tested at their own expense. Finally, if private doctors saw the free city tests as a direct lure of patients away from their own offices, animosity between the physicians and the Chicago Board of Health would only be exacerbated.

Thus, the existence of thirty-one free testing sites "aside from" the hundreds of private physicians' offices was a source of potential contention between Chicago's public and private medical communities. Moreover, while officials continued to praise the generosity and support of the private medical community, they also made it clear that cooperation was expected. In the first issue of the *Chicago V.D. Bulletin*, the monthly publication of the Public Health Service, Louis Schmidt challenged, in a tone that would hardly seem to endear him to private physicians (perhaps justifying Health Service concerns about Schmidt's ability to forge a good relationship with private medicine), "Any venereal disease clinician who is not a subscriber [to *Venereal Disease Information*] is not interested in the progress of venereal disease control and evidently does not expect to remain long in the service."[7]

Even stronger methods of persuasion were tried. Upon learning from a patient, investigator, or death certificate that a physician had failed to report any case of syphilis, gonorrhea, chancroid, or granuloma inguinale, O. C. Wenger would send a letter to that physician saying that he would "appreciate having an explanation" of this lapse and urging future "prompt" reporting. If this warning failed, physicians could be asked to report to the Health Department for a formal warning "that another failure [would] mean prosecution."[8] Wenger, however, observed, "Since it is quite difficult to convict a physician for failure to report under the present law, the advice of legal counsel concerning a new approach has been sought unofficially."[9]

The situation of Chicago's private physicians, then, was in some ways similar to that of Chicago's citizens, who were urged to volunteer for a program empowered to mandate some degree of participation. While Chicago's officials asked private physicians to cooperate with the program, there was also an undertone that belied the spirit of voluntarism. How far could the program or the Board of Health go in forcing cooperation? The initial strategy, put before Chicago's physicians early in 1937, was to invite responsibility and reward participation as much as possible. "One of the basic objectives of the Chicago Venereal Disease Control Program," Wenger outlined in his first annual report, "was to have all persons, sick and well, report to their own physician for a physical examination which would include the taking of a blood specimen for serological test."[10]

Although a private physician could not "ethically invite people to come into his office," a recommendation by the Health Department was wholly appropriate. Moreover, such an arrangement would be mutually beneficial:

> By this plan it was hoped not only to uncover venereal disease but, at the same time, to bring the physician into contact with a potential patient, since, during the examination he might uncover other illnesses or disabilities in the individual. By this method, also, the blood test would be obtained at practically no cost to the health department. Since anti-syphilitic drugs would be furnished for all cases uncovered, regardless of financial status, it would be clear profit to the physician to treat the cases uncovered who were able to pay a fee. The physicians agreed to send all indigent cases to the free clinic.[11]

The arrangement would profit private physicians, who appeared to agree. "All of this was explained to the medical groups repeatedly at various meetings and received their unqualified endorsement," Wenger concluded this section of his report, but the deliberately strong words *repeatedly* and *unqualified* sound almost like a set-up for succeeding sections.[12]

At the same time that the Public Health Service was offering financial incentives to physicians it embarked on two consecutive surveys. The first sought to determine how many doctors treated syphilis and gonorrhea. More than 80 percent of the 5,625 questionnaires were returned, and more than two thousand physicians reported that they did treat people for syphilis or gonorrhea. The respondents were next visited by a physician employed by the Illinois Department of Health, who reviewed with them the objectives of the Chicago Syphilis Control Program, distributed new forms for reporting cases and ordering drugs and laboratory services, and reminded them of the free services available through the city and state health departments.[13]

The first survey was deemed a success, with a 421 percent increase in cases of syphilis reported between March 1 and June 30, 1937, compared to that same period in 1936.[14] "War on Syphilis Spurs Reporting of Cases in City," the *Tribune* proudly announced.[15] A second story credited the antisyphilis campaign and free testing for bringing more people under treatment in the early stages of their disease.[16] By the time these statistics were compiled, though, a second, even more ambitious—but ultimately less satisfying—survey had begun. In mid-June, all of Chicago's private physicians, as well as all hospitals and clinics, were queried again, this time about all of each physician's patients (without names) under treatment for syphilis from March through June. For each patient, the physician was to indicate age, race, sex, marital status, ward of residence, date of initial visit, date of most recent visit, and stage of each patient's disease (congenital, early, latent, or late). About a month later, second letters were sent to physicians who had not responded, offering the assistance of one of thirty-five senior medical students, each of whom had taken a basic course in reading data on clinic records "at their own expense," and were being used primarily to extract similar data from hospital records.[17]

The response, Dr. Wenger reported, was "remarkable." Nearly 100 (99.6) percent of the private physicians completed the questionnaire, even if only to indicate that they were treating no patients for syphilis; the response from hospitals and clinics was "equally gratifying." The total number of patients under treatment for the four-month period was 19,071—23.7 percent by private physicians.[18] This news was both good and bad. The good news, which was the only news reported in the *Tribune*, was that about one-fourth of the cases reported were from private physicians, compared to only one in eleven of the cases reported in 1936.[19] The *Tribune*, however, did not distinguish between "reporting" via the questionnaire and reporting cases to the Board of Health. In actuality, the questionnaire gave the

Tribune its numbers for 1937, whereas official city numbers were the source of information for 1936.[20] The officers of the Chicago Board of Health discovered, to their great dismay, that although private physicians reported on the questionnaires that they had treated 4,515 patients from March to July, only 985 of those cases had been reported to the Board of Health. Even assuming that some of the 4,515 patients had been diagnosed earlier (rather than recent diagnoses that would have been reported during that period), statisticians still found that the number of patients reported for the eighteen months from January 1, 1936, to June 30, 1937, was only 40 percent of the numbers reported unofficially through the survey for four months in 1937.[21]

As early as February of that year, Ben Reitman had noted that only a small number of physicians were reporting cases of syphilis, not the response that had been anticipated.[22] Physicians did, however, report four times more patients and treatments for syphilis in 1937 than they had in 1936, probably due to the various efforts of the Board of Health to generate more tests, but leaders of the Chicago Syphilis Control Program found the results disappointing. "There are still many physicians not reporting any cases and others not reporting all cases," was their only conclusion, and they were forced to concede, "The plan to have physician[s] take serological tests [of their patients] has not worked out very well."[23]

Casting the Net

Although the *Tribune* continued to write glowingly about the increase in requests for syphilis tests and the contribution being made by private physicians to Chicago's antisyphilis effort, matters were not so rosy behind the scenes. Besides low reporting, the initial surge in testing was not sustained as the summer progressed. As early as November 1937, results were, in Reitman's words, "surprisingly low."[24] Although the Board of Health Laboratory ran 701, or 5.5 percent, more tests in October than September, private laboratories showed a concomitant drop in business because, officials surmised, more patients turned to the free clinics. Reitman concluded, with a mixture of uncertainty and disappointment, "It is difficult to explain why there hasn't been more blood tests as the poll activity has gotten a good start."[25] His disappointment grew the next month, when only 136 people who were tested at the City Hall Marriage Clinic came in response to the citywide letter of the spring. "It's a little too early to make the statement, but of the 100,000 letters sent out inviting people to get blood tests it is doubtful if more than 10,000 persons responded and 500 cases of syphilis were discovered."[26]

The next phase of the program would attempt to offset the low numbers. While continuing to seek the support of private physicians, the emphasis of the program was shifted to its own clinics. Nearly 100 percent of the cases indicated by responding hospitals and clinics on the second survey appeared to have been reported officially. Much of the credit to this high correspondence was given to the fact that the Municipal Clinic routinely reported every case.[27] The Municipal Social Hygiene Clinic (MSHC), run by the Board of Health, offered free testing and treatment to all comers. By shifting its strategy and urging people into city clinics rather than private offices and clinics, the program had a greater chance of ensuring ongoing, relative accuracy in reporting—and thus following—syphilis. It would, though, also be drawing many potential private patients into city clinics, at least for the initial test, risking the wrath of private medicine.

But it was a risk that seemed worth taking. By September, stories in the *Tribune* began placing new emphasis on free city services for the poor. "List 35 Clinics in City for Free Syphilis Tests" provided the addresses and hours of all free testing sites and mentioned only that private clinics and "many" private physicians would also give the test.[28] A human interest story told of a "nervous and melancholy" man who walked into a judge's chambers and "asked abruptly what crime he could commit that would insure him three years in prison." Judge Heller discovered that the man, from a small midwestern town, had contracted syphilis and was too poor to afford the treatments. He concluded that the best way to be cured was to enter a facility where treatment would be provided. Heller sent the man to Commissioner Bundesen himself, and the man was soon doing "odd jobs around the place to pay, in a measure, for his keep." Bundesen told reporters, "This instance shows the real value of Chicago's campaign to wipe out syphilis. Any syphilis sufferer who appeals to the board of health as this man did will receive sympathetic attention and information on what to do. Our facilities should have more use."[29]

Although the shift in strategy was relatively unheralded, it happened quickly; administrators of the Chicago Syphilis Control Program felt that the "physician's failure" to become an influential partner in the drive threatened to slow the momentum that had built during the summer. It became "imperative that specimens be collected before the public loses interest."[30] The shift from private to public services was presented, however, as a response to continually growing demands for testing. On December 15, the *Tribune* announced, in a burst of alliteration, "Survey Shows Syphilis Drive Barely Begun." In a presentation before the Chicago Medical Society, the newspaper reported that R. A. Vonderlehr, assistant surgeon general and chief of the Public Health Service's Division of Venereal Diseases, and Lida J. Usilton, a Health Service statistician, had announced that "only" one-third of the estimated forty-five thousand people with syphilis in Chicago were under treatment. Although private physicians were reporting more patients, only one in ten people with syphilis could afford treatment, the *Tribune* said, thus implying the need for other than private care.[31]

On the same page, a shorter article noted a proposal for a new city venereal disease clinic that could offer improved and expanded space for many who were unable to pay the costs of lengthy treatment.[32] About six weeks later, in early February 1938, the *Tribune* announced that twenty-seven special clinics would be open for five weeks, five days a week, at various hours. These clinics, staffed by four teams of physician and nurse, would provide "free and secret" syphilis testing to Works Progress Administration employees and people on the city's relief roles, but the rest of Chicago's citizenry was also "invited" to partake of the service.[33] Within a week, the clinics were running six days a week and had added a fifth team. The Chicago Syphilis Control Program announced a special five-week drive directed specifically at the thousands of people who had responded in the survey that they wanted blood tests. Three to six stations would be open every day. The *Tribune* once again listed the stations, their addresses, and the days that they would be open.[34]

The dragnet had begun in earnest. Borrowing the term from other public health campaigns in the United States, the aim of the syphilis dragnet was to reach even those people who might not be aware that they had venereal disease, the "unknown sources of infection."[35] For such a campaign to succeed, it would have to reach into all the streets and neighborhoods of Chicago and be seen as important and nonthreatening by those unknown, and often unsuspecting, sources. The early results of the new push suggested that the goal might be achieved: "3,078 Take Free Syphilis Test in Three Weeks," the *Tribune* counted at the end of the month.[36] And the numbers climbed even more the next week, as 5,254 more tests were given.[37] "Fourteen hundred blood specimens in one day," Reitman boasted on February 17. "The week of February 14, 1938, has been the banner week in the [Board of Health] Laboratory, as far as blood specimens are concerned." He credited "the new stations where the W.P.A. and the Relief clients go for their tests" for this increase, and he predicted that "if specimens keep pouring in at the rate they are so far this year, the Chicago Board of Health will be between 150,000 and 200,000 blood tests, in 1938."[38]

Business continued to accelerate, even in the long-standing clinics such as the MSHC. "When the director of the MSHC arrived at 8.35 A.M. there were one hundred patients waiting in the clinic room and fifty had already been treated and left," Reitman described. Those hundred patients, all men, were "a serious middle-aged group," whose clothes suggested that they were largely "working men or unemployed. Only 6 of them had white collars. Not half of them had neck-ties." He was impressed by their solemnity, describing them as "a silent, but not a sullen crowd. Everybody appeared serious and they took their place in the line without any rumpus, though a few tried to shove ahead." Each man was "given a number and told to line up. A well trained clerk speedily handed out the charts as soon as the patients told their number. The clerks were soft spoken, friendly and courteous. One patient who had a blood test three weeks ago was told that the record of his case could not be found. Several clerks attempted to find the record without success. The situation was intelligently explained to him and he consented to another blood test." Reitman reported a usually efficient, respectful staff that worked at delicate, intimate matters in less than perfect conditions with the dignified, health-minded poor. Even conflict, Reitman implied, was swiftly and kindly resolved. "The orderlies led the patients to the treatment rooms and in unbelieveably few minutes they came out of the treatment rooms, putting on their coats as they hurried to the elevator. Other patients were hurried to the social service room. The first 100 in an hour and the next 1500—the day was over."[39]

Although he applauded the staff's hard work and commiserated with its heavy workload, Reitman began to register some concerns about the ultimate effectiveness of the dragnet. For example, he worried about the completeness and representativeness of the results that the program was achieving. For one thing, "The difficulty of getting a large section of all the different social statuses of society in for a blood test grows no easier as the months progress," Reitman observed. He became increasingly aware of the underrepresentation of both Anglo-Americans and people from the wealthier classes: "They have got groups from the WPA and the Relief Stations tested but none from the Chicago Athletic Club or from the Real Estate Board or from the Banks. What we need now is to find out how many persons belonging to the Chicago Club, the Opera Club, the University Club, the Police Association and the Rotary, and the Kiwanis have syphilis."[40] Certainly, in inviting people to avail themselves of free services in spartan city facilities, most of which were located in poorer sections of town, the Chicago Syphilis Control Program would have appealed to a different clientele from that which would have sought private physicians. Statistics from either group alone, though, might not be representative of the whole of Chicago's population.

Reitman also expressed doubts about "official" estimates that 10 percent of the country's population was infected. As early as July 1937, he was noting lower numbers of new cases than expected.[41] As project tabulations emerged over the ensuing year, the rate of positives was closer to 6 percent, and Reitman was beginning to wonder whether even that figure was too high (even though he continued to worry that the number of people being tested was too low). He "desired to point out once more that the law of probabilities is that in the late syphilis dragnet we have caught the biggest fish (the largest amount of old syphilitic cases). The syphilis dragnet in the future will get less and less positive diagnostic cases."[42]

In these early months of the syphilis dragnet, however, the daily work was lively and the mood often optimistic. In the late winter of 1938, it seemed that one of the first goals of the Chicago Syphilis Control Program—to test all citizens of Chicago—was about to be realized after near disaster through the program's failed partnership with the city's private medical community. The five weeks' operation of special city clinics was extended to seven weeks, and, at the end of that period, the officers of the program decided to keep the clinics in operation nine weeks more: "36 Permanent Syphilis Testing Offices to Open; Intensive Public Program Undertaken by City," cried the *Tribune* on April 4, 1938. In one week, six teams comprised of a city physician, nurse, and WPA clerk would, by "public demand," staff "health welfare stations," nineteen of which were located in the district field stations of the Chicago city parks. The addresses of the stations and days of their operations were printed once more.[43]

That the Chicago Syphilis Control Program was picking up work intended for private physicians was made clear in the program's first annual report: "Since the private physicians have been given a year to make this survey and failed to do so satisfactorily, special serological crews have been organized to do this work."[44] The *Tribune* reported the use of some city employees, but Wenger's account clarifies that the "serological crews" were all on the payroll of the Works Progress Administration. The thirty-six stations, twenty of which were "located in permanent, established infant welfare stations" and sixteen of which were "in park field houses, school buildings, and other public establishments," would be added to "as needed." Each of the six crews visited a different station during its six-day work week. The stations were open from 3 to 8 P.M.[45] All a person had to do was walk in.

"Bring Them In"

From the start of the expanded dragnet program, and despite any other apprehensions he might have expressed, Reitman entered into its activities with an energy and optimism that surpassed even his usual exuberance. "Bring them in from the fields of sin," he sang in his reports, adapting the old hymn to the call of public health. He visited both old and new testing sites almost daily, many of which were located in neighborhoods he knew well. Part of his job was to distribute cards around neighborhoods before the station in that area opened, but he soon found a much larger venue for his particular interests and skills. "A hasty visit to three south side syphilis drag-net stations was revealing," he reported soon after the full program began. "The station at 31st and State Streets is an ideal location and business was good. We took the liberty of taking Dr. Simpson up to the corner saloon and introduced him to a large crowd of men and women gathered there at six o'clock in the evening. We said, 'Boys, this is Dr. Simpson, who is at the station down the street. You can go over there and get a blood test free.'"[46]

Whenever possible, Reitman sought out groups of people whom he considered most in need of testing—or tolerant of his solicitations. He deemed local bartenders and pharmacists to be the best sources of information on the prevalence of syphilis and gonorrhea and the openness of a neighborhood's citizens to testing. Reitman's tour continued. He reported, "At 9016 Cottage Grove Avenue business was slow. The neighborhood was sparsely populated. The street was quiet. The station was well equipped and everyone was ready to work, but there hadn't been a dozen patients there that day. At the corner drugstore the proprietor didn't seem to know that the station was there and functioning. It is a fine residential neighborhood but it is going to be a difficult task to get the quota of blood tests there, but industry is always rewarded and the blood tests will come in." Location was of constant concern to Reitman. Busy locations with heavy foot traffic at leisure times were best. Parks were fine if they were well populated and there wasn't a ball game to hold people's attention. Residential streets or secluded parks were the worst unless the latter contained a frequently used bath house (then a major part of Chicago's numerous neighborhood parks) that had a cooperative manager.

Reitman's rounds took him next to 437 West 119th Street, the site of a clinic that would open the next day. "It is an ideal location," he noted:

> Just across the street from the station there is at present a large carnival with hurdy-gurdies and ferris wheels etc. The entire neighborhood was out enjoying themselves. Cards were passed to hundreds of merrymakers and everyone was invited: "Boys, tomorrow you have a chance to have your blood tested free. Across the street at 437 119th Street there will be a doctor and nurses from three to eight P.M. Be sure to come and have your blood tested. Most of you are all right, but if you are sick you ought to know it. It is the intelligent manly thing to do." Cards were also distributed in taverns, drugstores, garages and filling stations, wherever there were a few people.[47]

He was tireless in his recruiting. He was doing the kind of work he most enjoyed: meeting and talking to all kinds of people, urging, joking, questioning, teaching. "Wherever there were a few people" had always been a place where Reitman liked to be. The job seems to have been made specially for him—and probably was.

In the first weeks of the dragnet, Reitman took Schmidt and Wenger to meet hoboes in the McCoy Hotel, even poorer transient men and women in the Cake (flop) House, prostitutes at Emily Marshall's famous house, and the workers and customers of several taverns on West Madison Street.[48] Before long, Reitman was regularly bringing members of field teams along as he made his rounds of the dragnet stations, teaching them about each neighborhood. For example, on April 25 he reported, "[Eck]hardt Park, at 1400 West Chicago Avenue, has a fine drag-net station. With Bantus [the physician assigned to the station] we covered the district, and stopped at all the taverns, drugstores, poolrooms, cigar stores, and other places along Ashland Avenue, Chicago Avenue, Milwaukee and down to Halsted Street. It was Saturday night and business was good. Many of the taverns were crowded and we had opportunities to make our little speech and distribute cards."[49] Although few tests resulted from these visits, Reitman felt them beneficial for educating the public and studying the local mood. That night, he found, "We met with very little hostility but the majority of men listened indifferently, and then went back to their beer. We noticed that many of the men seemed to be buying liquor by the bottle."

After their efforts in the poolrooms and drugstores around Eckhardt,

Reitman and Dr. Bantus returned to the station, where Reitman took up a position on the street and began "roping," working the crowd. For an hour he stood in front of the dragnet station and stopped every passerby. "Wouldn't you like to have a free blood test tonight? This is a fine chance to learn whether you have syphilis or not. Step right inside. The test will only take a minute. There's a doctor and nurse inside. It won't take but a minute. Come inside and get a blood test." Reitman estimated that about 10 percent of people thus approached agreed to be tested—all but one of them men. One woman, he recounted, "cursed us for daring to suggest that she might have syphilis. But the men took it goodnaturedly and we were convinced that if there were more passersby there would have been more customers."[50]

Thus was fashioned Ben Reitman's new, unofficial job with the syphilis testing stations, work clearly addressing his charge to "popularize" the work of the Chicago Syphilis Control Program.[51] But even roping was only a modest part of what Reitman felt could be done to bring people into field stations, and he frequently made suggestions for attracting more people. He proposed that officials "dramatize the stations by a series of maneuvers and theatricals"; develop large billboards and posters to advertise the presence of the stations; canvass homes, stores, and taverns; offer prizes to stations that took the most tests; use neighborhood children as recruiters; relocate some of the poorly placed stations so they would be more visible and accessible; and, finally, "have a brass band, music, a demonstration, or a movie."[52]

Reitman continued to canvass stores and taverns as much as he could but discovered that he was actually abashed when groups of children attached themselves to him and began "roping" in imitation of him.[53] The band and theatricals never materialized as Reitman described them, although a play, *Spirochete*, by Arnold Sungaard, ran at Chicago's Blackstone Theater for several weeks. Part of the Federal Theater, a New Deal program that aided playwrights and performers, the play documented "the story of syphilis from the time of Columbus to present day problems."[54] Booths were set up in the lobby of the theater, and health officials performed free blood tests before and after the show and during intermission. At one point, Wenger (surpassing even Reitman in outrageous invention) suggested that "Thursday night performances be free to all patients who have syphilis in Chicago," but there is no confirmation that his idea was ever put into practice.[55]

Prizes for top testing stations were not instigated, but stations soon were identifiable by a sandwich-board sign. Another federal program, the Federal Art Project, was enlisted to design the signs, four by three feet, which were placed in windows or set outside the clinics on A-hinges. Foot-high, inch-wide lettering announced "Free Blood Tests."[56] Because Wenger does not credit the signs to any particular person, Reitman's contribution cannot be judged. His comments in his reports may have been part of general conversation and discussions going on around bringing business to the clinics, but his suggestions at least indicate that he was often privy to leaders' concerns about the project. And, whoever suggested them, the signs seemed to help. Wenger concluded his first annual report with the following anecdote: "At one serological station no applicants for tests appeared until a large art project sign, announcing FREE BLOOD TESTS WITHIN, was placed outside. Then, in a few minutes' time, there were 50 applicants."[57]

Formal relocation of stations did not occur, but two departures from the established stations are documented. One Saturday in June, finding business especially slow because of a baseball game in the park, "Dr Harris, and Dr Jenkins took their outfit and a little table over to the ball grounds and started to take bloods. They had only 50 syringes and no facilities out there so they quickly took 50 bloodtests."[58] The other departure was even more daring. Driving along Madison Street one day around this same time, Dr. Wenger and his staff (it is not clear whether Reitman was among them then although he was later) spied "a little shoe repairing shop at 1723 west Madison. It is a very small shop, a few shoeshining chairs and some shoe-repairing machinery." They asked the Italian proprietor "if he would permit the board of health to open a bloodtesting station. He didn't know what it was all about, but they looked like police to him, but since it was official business, no bootlegging or gambling attached to it, he gave permission."

A few days later, Dr. Scott, a nurse, and Reitman returned to the shop. "The shoemaker was the first client, he sat down on the stool and Dr Scott put a rubber around his arm and started to draw blood and the nurse started to write names, the barker started to yell, 'Come right in and get a bloodtest' and 50 persons came in and 50 persons got a bloodtest and that's all there was to it." Reitman was delighted with the results and with the setting, beside a tavern and below "some kind of a 'joint,'" which yielded a steady traffic and several blood tests. He concluded, "[It's] clear that the problem of getting bloods in Chicago is simple: all one has to [do is go] where the crowd is. This business of sitting in a station, waiting for patients [should] be a thing of the past."[59]

As much as Reitman praised the new approach, it didn't happen very often. It was a rare health team—and a rare proprietor—who would agree to such an unorthodox arrangement. Reitman's reports, however, often noted the presence of Dr. Scott, supervisor of the dragnet stations; Dr. O. C. Wenger; and Dr. G. G. Taylor, the director of the Venereal Disease Program of the Chicago Board of Health. Wenger was not new to this kind of field work. Less than a decade earlier, Thomas Parran, then head of the Division of Venereal Diseases in the Public Health Service, had taken Wenger from his assignment running the Public Health Venereal Disease Clinic for the indigent in Hot Springs, Arkansas, and sent him to Mississippi, where he set up and ran a "syphilis demonstration project" that served not only to bring treatment for syphilis to poor, largely black southerners but also to be "a springboard for establishing federal leadership and control over a national campaign to eradicate syphilis." From there, Wenger went to Macon County, Alabama, where he set up another demonstration, which led to the creation of the Tuskegee Study. In all of these programs, he worked daily with patients, government officials at the local and national levels, and influential philanthropists.[60] Certainly, Reitman's approach to recruiting volunteers for blood tests would have struck a sympathetic chord with this veteran campaigner.

The renewed effort of the Chicago Syphilis Control Program seemed, at least initially, to work. Between January and May 1938, 88,714 tests were given. Even allowing some of these to be repeat tests, the number marked nearly a threefold increase over the 28,047 tests given during the same period in 1937.[61] Moreover, there was a marked increase in the number of people being brought under treatment. In the month of May, for example, 1,694 cases were discovered and treated, compared to 1,065 during May the previous year.[62] Nearly every *Tribune* headline boasted a number—of tests taken, positives uncovered, special groups being tested. Despite all the hopefulness, though, the dragnets did not last long. By the end of May 1938, after only about four months of business, activity began to lag.

"Now We Need Results"

It was rainy in Chicago during the summer of 1938, one of the wettest summers on record, according to Reitman. Whether or not the weather was totally to blame, some dismal afternoons and evenings were spent at certain dragnet stations. Other days were a flurry of activity. Reitman's optimism rose and fell, but his conviction about the value of the Chicago Syphilis Control Program and the field stations themselves never faltered. Health was a right to Reitman. In the fight against venereal disease, all men and women were equal and to be respected equally. Thus, his reports of the activities of the syphilis dragnet stations not only yield a vivid picture of this phase of the activities of the program, but also embody the spirit of the program's purpose as well as a few of Reitman's own.

"Let us be up and at 'em," Ben Reitman urged his bosses in May 1938, referring to the disappointingly small turn-out to the neighborhood dragnet stations. "There has been plenty of speculating and theorizing and advice given and suggestions made already. . . . We have had enough talk. Now we need results. We know that if a couple of solicitors stand in front of the station we can get from ten to fifty people in for a blood test. So, what more is there to say?"[63] Although he characteristically still had more to say, he followed his own advice about putting word to deed. He was often able to improve business, but at times even his best efforts produced a small return.

By early June, Reitman's reports already sounded a note of uncertainty about the future of the field stations. "Yes, why should a dragnet crew sit around a station waiting for non-existent patients," Reitman agreed, but he still went on to argue against closing the stations.[64] In the next report, he reemphasized the advantage of a roper but admitted the added cost of such a person. "Whenever the 'ropers' have been there the number of tests taken have gone up to 50 and when they were not there the number dropped to a baker's dozen. . . . Dr. Scott is a most efficient 'roper,' but he is a busy man and has the supervision of 30 stations and hasn't time to stay at any one station. Repeatedly we have urged that the clerks at these stations be trained to 'rope,' but on Wednesday night there was no clerk present; there wasn't even a V-shaped sign."[65]

Reitman's concern about both the continued poor showing in many stations and his belief that the stations were, nevertheless, valuable became frequent themes of his reports as the summer of 1938 wore on. His tone grew even more urgent as continued operation of several stations fell into jeopardy. "We didn't have long to work," he reported on June 30, "but we did get a dozen blood tests at the park and 7 at the station, making a total of 19, which was a poor showing. But next week it will be easy to get more. We shouted 'get a free bloodtest' into the ears of nearly 1000 persons and they're not going to forget it very quickly. Stations 81, 59, 67, 72, and 84 should not be given up. If more effort and imagination is thrown into these they will yield satisfying results."[66] Reitman's advice, however, went unheeded. Over the next few months, dragnet stations were quietly but steadily closed.

The stations proved to be more costly than other alternatives. One day, during a ride with Lawrence Linck, Reitman was told, "It appears that the cheapest, best, and most satisfactory method of doing bloodtests is to bloodtest the employees in industry, or in schools, or large organizations like the YMCAS, Boy's Clubs, etc. Here bloodtests can be gotten quickly and cheaply. The Cook County Hospital and the City Hall dragnet stations are the least expensive to operate."[67] Reitman, however, saw the neighborhood stations as serving an educational function equally as valuable as their medical one. "The fact that hundreds of cases of syphilis have been diagnosed justifies all of the expenses entailed," he insisted. "The syphilis dragnet stations, the signs on the street, the efforts of Chicago to bloodtest its citizens has made Chicago syphilis conscious." Rather than reduce the number of neighborhood dragnets, he urged that "under no consideration will any of the stations be given up. More stations should be opened. Let us set a goal of 1,000,000 bloodtests in 1938 and 1,500,000 in 1939."[68]

When the Municipal Social Hygiene Clinic moved into spacious new offices at 26th and Wabash in late June, it installed the tabulating equipment of the Chicago Syphilis Control Project in its basement.[69] By August a special tabulating system was producing statistics that differentiated among the various stages of syphilis, and on August 5, the *Tribune* reported that although more people in Chicago than in past years were being treated for syphilis, fewer of those cases were infectious.[70] Reitman's earlier prediction that the dragnets would mostly uncover cases of "old" syphilis was finally confirmed. Still, he felt that there was much work yet to be done. "It is absolutely true that for the first time in the history of Chicago that the 'pools' of old syphilis are being drained," he acknowledged, but cautioned, "Of the 9,000 cases of syphilis reported from all sources in the first seven months if it had not been for the syphilis drive and the dragnet stations, fully 5,000 of these cases would have remained in the cesspools and festered and polluted the population. Let us hope that we will develop a better methodology of draining the pool."[71]

Reitman's "better methodology" was a twofold plan, both parts of which turned on his long-held beliefs that sexual expression was a part of human nature and that syphilis was only one of the many diseases that should be handled through public programs of education, detection, treatment, and prevention. From the beginning of his reports, he had spoken of the prevention of venereal disease, urging as early as September 1937, "But nowhere in the city is there a place for the man or woman who has been exposed to venereal disease to get prophylactic treatment."[72] He recommended the creation of prophylactic stations or prevention centers, a recommendation he repeated with growing frequency over the next few months. In early September 1938, with all but ten of the dragnet stations closed, Reitman directly asked Dr. Taylor about his position on the subject. Taylor responded, "We have a job to do and we are trying to do it honestly and well. Our business is to get all the bloodtests we can, and to do everything in our power to aid the syphilis drive, but our business is not prophylaxis. Of course we believe in prophylaxis just as much as you but that isn't our job." Reitman was disgruntled. "It was a very nice Dutch lunch," he concluded his report. "We left the Italian restaurant muttering 'How long, Oh Lord, how long?' "[73]

The second part of Reitman's plan put syphilis within a broader context of widespread, government-sponsored health care for the poor. Seeing the treatment of syphilis as only one aspect of the work of the Board of Health, in mid-August Reitman urged, "Instead of closing the dragnet stations, they should be enlarged and increased. Oh, that we had the pen of a prophet to make clear that the MSHC and the dragnet stations are not treatment clinics nor even diagnostic centers: they are health community centers to carry the glad tidings of health, to prevent diseases. . . . We could do physical examinations, urine analysis, sputum analysis, vaccination, serum injections, and so many other things. Is there one who can think in terms of the health of the community or the welfare of the people?"[74]

Closing of the stations continued, however, and on March 3, 1939, Dr. Linck wrote a memo to Marvin B. Osterman, assistant director of the program, ordering all but two of those remaining stations closed. Upon receiving a copy of that memo, Reitman immediately wrote to Linck to make a final plea, not just for the dragnet stations but seemingly for the entire Syphilis Control Program:

> From afar off, we have followed you. The illness of your staff, your own flu, your location without a lease, and the cry for more blood. We understand it all, and we reach out our hand to comfort you, but we are sorry that you are unable to get more joy out of the day's work, and more comfort out of your friends, and more determination to carry on. You mustn't think of giving up the ship. Hard work is never easy, and carrying on an intelligent syphilis control program requires a great deal of brains and strength and faith. I think it would be a serious mistake for you to make any change now, because you haven't made good, because you have permitted yourself to become irritated and disgusted. Stay right where you are.

Last month, as you will see by the record, there was a decrease in 4,000 syphilis cases reported and there were less than 800 cases of gonorrhea reported. The number of bloodtests reached an amazing low. I am not blaming you because we haven't got more syphilis.

Keep sweet. Let's dine together soon.[75]

CHAPTER SIX

Net Results

In the public's eye, the early months of Chicago's antisyphilis campaign had been fast-paced and eventful. The imminence of mass testing had been a source of civic debate and even spectacle as the stories of the citywide survey and the drop in marriages in Illinois continued to provide lively copy to newspapers well into 1938. Chicagoans were repeatedly told that they were flocking to testing stations in record numbers; those who were not doing so were chided by their commissioner of health, their ministers, and, hopefully, their consciences. While Chicago was the center of much federal money and effort, it was not alone in its campaign, however. By the end of 1937, the Public Health Service, following its offer to send special consultants to all states that requested help, had begun syphilis control programs across the country.

Alabama, which had previously provided free drugs only to indigent patients, now followed Chicago's example and began furnishing free drugs to physicians for all their patients.[1] In October, the *Tribune* announced new or expanded programs in Omaha, St. Louis, and Fort Wayne.[2] Michigan, expanding its municipal laboratory after experiencing an increase in business with the enactment of its own marriage law, announced its free drug program in early November, and New York City and New Jersey announced free testing around that same time.[3] Ben Reitman reported to his bosses that the syphilis control program in San Francisco included a plan for prophylaxis, but he offered no further details.[4] Private money also went to fight syphilis, as the Zachery Smith Reynolds Foundation gave all of its annual budget of $7 million to the antisyphilis campaign in North Carolina.[5]

Chicago, however, still continued to represent a major investment of the Health Service in attention as well as dollars. One reason for this importance lay in activities going on behind the scenes. To the Health Service, the information about syphilis that the Chicago program could generate was equally as important as bringing all who had syphilis into treatment. Epidemiology was an important tool in the armamentarium of public health during the 1930s. In Chicago, such work had begun with the two surveys of physicians and medical facilities in the spring and summer of 1937. The second survey, which hoped to record the age, sex, and dates of visits of all people being treated for syphilis, was to provide the basis for a broader project: an ongoing, detailed database from which statisticians could study the demographics of syphilis in Chicago and design other programs across the country. The information about amounts, types, and dates of treatment would be augmented by names and addresses of a patient's sexual contacts as far as that was possible. The forms the Public Health Service created would be analyzed, revised where needed, and then become the standard forms by which all physicians, hospitals, and clinics would report cases of syphilis.

Thus, behind the novelty of the A-hinged signs or the ambitiousness of mass testing, there was another, perhaps less novel but even more sizable, project involving hundreds of people. Their job was to catalog, count, and interpret the flood of numbers pouring from the laboratories, offices, and clinics across the city. By April 1938, besides the medical students recruited to cull information from hospital records, ninety-nine Works Progress Administration employees found work as statisticians, and thirty-six more were assigned to the serological crews. Another 155 people in the WPA and 23 in Social Security were working in the Municipal Clinic.[6] By the end of 1939, those numbers would be even larger, with 605 employees, 480 of them on the payroll of the WPA, working as health professionals, clerks for the statistical and secretarial tasks, educators and research assistants, and in countless other positions in this extensive undertaking.[7]

The statistical work of the program raised special challenges. Besides the logistical problems inherent in mass testing, other problems lay in how to record and interpret the information gathered. Second, officials of the program had to consider how to report that information to government officials, to Chicago's medical community, and to Chicago's citizens. Not all numbers, after all, can be sensationalized with pictures of people crowding Chicago's City Hall for a marriage license—nor would officials always want such attention. Third, even counting numbers or reading a slide of a blood sample takes place in an environment fraught with political and fiscal tensions. The inseparability of the problems raised by the very acts of gathering, interpreting, and making decisions based upon statistics became evident to the members of the Chicago Syphilis Control Program in the early months of voluntary testing.

"Standardize All Laboratories"

The great detail in reporting that the program now demanded of physicians also required a new rigor in laboratory methods and procedures. Besides seeking more reliable positive readings, public health officers also wanted to be able to draw finer distinctions within a positive reading, as epidemiologists sought demographic information about the various stages of syphilis. Inseparable from generating statistics, therefore, was generating accurate, useful laboratory results. At the time the Chicago Syphilis Control Program began, approximately 150 serological laboratories, both publicly and privately owned, were operating in the city, with no standard forms or protocols uniting them. When the Public Health Service offered to provide free antigen to laboratories testing for syphilis, it stipulated that any laboratory that received the antigen would have to comply with standards for operation. These standards would be written by the Illinois Department of Health, which would be responsible for distributing Health Service funds and drugs within the state.[8] (As part of his work for the program, Reuben Kahn himself had agreed to write the guidelines for the laboratories so that his test would be used properly.)[9] By April 1938, ninety-three laboratories had applied to the State Department of Health for approval.[10]

Because the state set the laboratories' standards, and because federal money for testing would be coming through the state Department of Public Health, Dr. Frank Jirka's office in Springfield was the obvious one to coordinate the work of Chicago's medical laboratories. Plans to that effect were, in fact, supposed to have been undertaken late in 1936.[11] In March 1938, however, Public Health official David C. Elliott was dismayed to discover that the intended site of a new state laboratory in Chicago was in total disarray, due in part to political tangles surrounding Jirka's resignation nearly a year earlier. The Chicago Board of Health's laboratory, however, was operating in its new quarters and offered itself "prepared to standardize all laboratories in Chicago." Elliott conferred with Kahn and, with the agreement of Acting Director of Public Health for Illinois Dr. A. C.

Baxter, "a full approval order was issued and notices prepared and sent to 225 laboratories throughout the State." Beginning March 1, 1938, the Chicago Board of Health Laboratory would administer, not just across Chicago but statewide, all the free syphilis antigen for the syphilis tests. To receive the antigen, each laboratory would have to adopt standards derived by Dr. Kahn and pass inspection on their achievement of those standards. Elliott noted that Michigan (Kahn's own state) was the only other state whose laboratories operated uniformly statewide.[12] Chicago's plan proceeded quickly and smoothly, and by late September, after one of Elliott's visits to Chicago's and Illinois' laboratories, Dr. R. A. Vonderlehr commented, "It is most encouraging to note that a large number of private laboratories are now participating in the serologic evaluation work conducted by the State Health Department."[13]

The Chicago Syphilis Control Program was established at a time when tensions between Chicago and Springfield were at an all-time high. The candidacy of Governor Henry Horner, a Chicagoan, had been strongly supported in 1932 by Edward Kelly, then chief engineer for the Chicago Sanitary District, a powerful position within the city's especially powerful political machine. When Mayor Anthony Cermak died from a gunshot presumably intended for President-elect Franklin Roosevelt, Horner agreed to Democratic boss Patrick A. Nash's choice of Ed Kelly for mayor. The relationship between Kelly and Horner began amicably, but eroded in following years as Horner made a number of decisions that did not favor Chicago (or the machine politics of Nash and Kelly). Some of the antagonism arose from Horner's charges that Kelly and his men were using "WPA jobs for their own . . . selfish ends."[14]

One of the more amusing confrontations arose in 1935, when it was suggested to Horner that the Illinois Department of Public Health distribute statewide Herman Bundesen's popular baby books, which were already given to all new mothers in Chicago and sold to insurance companies. Frank Jirka, however, convinced Horner that they both would be better served if the state wrote its own book with their (Jirka's and Horner's) names on it. Bundesen complained that this book would compete unfairly with Chicago's largess, so a promise was made not to send Springfield's books to Chicago. The venture ultimately proved too costly, however, and books were sent only to local health departments and hospitals. Nevertheless, feathers were ruffled.[15]

Competition between Kelly and Horner occurred at every turn, but it was bigger issues than baby books that led Kelly to run a candidate against

Horner in the gubernatorial primary in 1936. So great was Kelly's standing with Roosevelt that the national party swung its support wholeheartedly behind Kelly, gravely wounding and so angering Horner that he launched a fight that few people had expected. His opponent—Herman Bundesen. Thomas B. Littlewood suggests that Kelly tapped Bundesen because he felt that Bundesen would *not* beat his Republican opponent—that he believed a Republican governor was actually less of a political threat than a governor from his own party. Bundesen, in his spats and monocle, ran a comical campaign, annoying rather than inspiring influential backers with his penchants for serving them glasses of milk and taking them on long walks. To demonstrate his vigorous health, he would hurdle a courtroom railing and perform a few calisthenics before speaking. His advocacy of milk rang hollow to many of the dairy farmers whose herds had been ordered destroyed when Bundesen put through his clean milk ordinances several years before.[16]

Horner won the election, but his enmity toward Kelly, and vice versa, was complete. "My biggest mistake," he once said, "was when I made Kelly mayor of Chicago. I will tear off his mask and with the people's help I will purge Illinois of bossism and sinister influence."[17] Such were the events taking place while the Chicago Syphilis Control Program and its various activities were being organized. To say the times were not the most favorable for state-city collaboration would be gross understatement, however much such political goings-on were not (and still are not) unusual.

The new laboratory for the Board of Health was part of a new Municipal Social Hygiene Clinic, a converted high school at Wabash and 26th Street, whose acquisition and renovation the *Tribune* followed for nearly five months. It was a story worthy of coverage; the old MSHC had struggled valiantly to keep up with new, expanding demands in an old, limited space. Originally designed to treat three hundred to five hundred patients a day, it handled a record attendance of more than 1,700 on September 7. "The remarkable feat of treating 1752 patients on the 13th floor of the Police Bldg. deserves commendation, for it is not a simple thing to interview, examine and treat 700 women and a thousand men," Ben Reitman reported. "Without any rush or bustle, without any shouting or evidence of over-crowding, the Chief of the Division, with the skill of a gener[al] under fire, managed the clinic and took care of every patient, many of whom had peculiar needs. The enthusiasm and dexterity of the staff, the clinic nurses and doctors and orderlies, was a joy to behold."[18] The surge in tests also resulted in a corresponding increase in urine samples and blood smears sent to the laboratories. In the first weeks of September and even earlier, Reitman reported record numbers of blood samples arriving at the Board of Health Laboratory.[19] Both the clinic and the laboratory were under pressure. A new facility was sorely needed.

Money was also a problem. After trying unsuccessfully to win additional WPA funds to help them with their project, the Board of Health turned to the City Council.[20] Territorial quibbles and attempted forays into other agencies' budgets were common during the years of Chicago's close work with the Public Health Service, as noted in Horner's charges about Kelly's use of WPA jobs. Because of past or ongoing tensions between state and city offices, however, issues of boundaries in the work of the Chicago Syphilis Control Program, which spanned all levels of government, were often particularly sensitive. Given such delicacy, the reorganization of the municipal clinic and laboratory was completed relatively easily and satisfactorily.

"A Very Ingenious System"

Most of the visible work of the program—and most of the rhetoric around it—came from the Chicago Board of Health and other city officials. The acts of calling citizens to be tested and opening testing stations were designed to create a popular referendum for the "war on syphilis," a term health officials and journalists used deliberately to characterize syphilis as a tangible enemy that was formidable but not invincible. The enemy would be engaged through individual testing and treatment. The Public Health Service, however, saw itself as engaged in a different type of campaign, an epidemiological one, in which numbers and machines would take the measure of syphilis. Statisticians would study the patterns that the disease created as it spread, and so outsmart it.

From the start, the Health Service strove to keep this project separate from the physical, medical work in the dragnet stations and medical offices. "The general plan was to disassociate this [popular] referendum from the statistical work as much as possible," Elliott noted in August, just before the official opening of the clinics for mass testing.[21] The program's title actually referred to the full activities of all of the state, local, and national organizations involved in testing and treating syphilis, activities primarily promoted and orchestrated by the Chicago Board of Health. The statistical operations were controlled by the Chicago Syphilis Control *Project*, which, although usually included (as it is here) in general discussions of the Chicago Syphilis Control *Program*, was actually a distinct unit of the Works Progress Administration. The project served both the Public Health Service, which paid its salaries, and the administrators of the broader program. "The organization that I am with, the Chicago Syphilis Control Project," Director Lawrence J. Linck said, "[has] acted as a service unit, as the clerical staff which serves the program, in a sense occupied the position of a vehicle for operations for the entire program. Those in charge of directing the program legally have utilized our services."[22]

This undertaking, too, was begun only after some quick rethinking of the abilities and resources of city and state. The Health Service used statistics that came from the individual state health departments, where the federal service had begun automating tabulating systems during the second half of the 1930s. In 1937, Illinois had been sent an IBM system—called the Central Tabulating Unit—to support the new work in syphilis control, but on visiting Springfield in March 1938, Elliott discovered that the machine was still in its delivery crates. It had only recently been moved from storage at the state fairgrounds to a private storage company at the request of IBM—but at the expense of the Public Health Service. IBM, irritated by that fact that its equipment sat idle, asked to send it to a state where it would be put to use.

Not only had the tabulator not been installed, but data were not being prepared in a way that the unit could use when it finally was in operation. "*Not a single 1937 venereal disease report* had been punched for tabulation," Elliott reported. The state had been preparing all of Chicago's reports by hand since the Syphilis Control Program began.[23] Consequently, the statistical work was shifted to Chicago and run in conjunction with the Chicago Syphilis Control Program and the Chicago Board of Health, independent of the state office, similar to the shift in responsibility for the distribution of syphilis antigen. A few days after Elliott's visits to Springfield and Chicago, the *Tribune* announced that WPA workers had begun punching tabulation cards for the cases under treatment for syphilis in thirty public and private clinics. Information would be updated daily from the clinics' reports.[24]

"Install Machine" the headline read, the politics behind this development quietly glossed over (and perhaps the members of the press never knew that a change of plans had occurred). But what a machine it was. The Central Tabulating Unit was the only machine supplied through the Public Health Service for the exclusive purpose of recording information about syphilis. Thus, it held great promise for an in-depth epidemiological study. The machine was the star of the Chicago Syphilis Control Project, which compiled its own reports and made available to the program statisticians and state-of-the-art statistical facilities. Its use put the program on the cutting edge of public health studies.

"A very ingenious system was devised and set up in Chicago—the Central Tabulating Unit; that is the name for it," Linck announced with great flourish and detail. The unit, officially named "A Mechanical System of Morbidity Reporting on Treatment-Progress and Control of Venereal Disease," would help state, city, and federal health agencies "tabulate better the true facts of the disease with reference to persons under treatment." Using a summary form for each patient at every clinic, "senior medical statistical workers" entered information in coded numbers for that person's gender, race, disease, age on admission, and date of infection (in months). The codes were then transferred to a "summary punch card," which, in turn, was assigned a serial number. Information for each patient could be obtained through that number or tallied with all other entries carrying the same code number. The cards, Linck said, could be run at a rate of four hundred a minute, a feat heretofore unheard of. Linck explained the mechanical and mathematical genius of the system:

> An electrical contact with the contact roller of the machine then throws the card into a certain slot in this machine. The electrical contacts are set for whatever column on the card you want. For instance, the machine can be set up so that a sheet of paper lists the case data all the way across. This gives to the clinic director a complete picture of his clinic population every month. . . . [It] tells him how many laboratory examinations he has made for the month, how many patients treated in the various stages of the disease—that is, primary, etc. How many in the various sex, age and disease classifications at the beginning of the month, and how many at the end of the month. It tells how many did not come in for treatment, how many lapsed for 30, 60 or more days, how many were assigned on rest [the period between regimens of chemotherapy].

The columns of numbers that could be so quickly generated were invaluable for tabulating the numbers of people tested and the results of those tests. Numbers could be sorted by people's age, race, or political ward and be studied by Public Health statisticians to determine such things as which age, sex, and racial groups had the highest incidence of reported (as opposed to absolute totals of) syphilis; how quickly syphilis moved from infectious to noninfectious stages; and what additional tests or information would help them understand the patterns of the spread of syphilis even more fully.

If such information was not of immediate relevance to many of Chicago's physicians, Linck went to great pains to point out that it also had practical use in the treatment of individual patients. "We can show you," he continued, "if you want to know after three months, how many treatments the patient has had. You want to know, Are we making progress in his treatment?" In another instance, "The physician can direct his clinician to pay special attention to certain cases, so that when they come in for treatment, if they have been having arsenical reactions, their record is flashed, so that they will not be given arsenical therapy."

The information on the cards would also benefit public health officials, who could follow up on patients and their sexual contacts. Linck cited the example of "case No. 2949J," a man who was "at the mercy of circumstances such as clinics which are so busy he has to stand in line, though in most clinics he gets fairly routine treatment. He stands and gets an arm or hip injection. The clinician on the treatment end of the line is not in charge of the clinic and has no particular opportunity to talk to this patient as an individual, has no opportunity to know whether the patient has painful reactions." Linck observed the unfortunate necessity of such impersonal treatment in some of the busy free clinics, but he offered the example to demonstrate another feature of the Central Tabulating Unit. Five of 2949J's six sexual contacts had been examined, and three of them had been put under treatment. The remaining unreached person was noted as such on monthly reports "until that case is either discharged or that person is located," either by the clinic or the special staff on the Board of Health. Linck's description clearly posited that the Tabulating Unit would have multiple purposes for statisticians, public health officers, and clinicians.

Where Linck's proud description of the Central Tabulating Unit stressed its efficiency and broad usefulness, Reitman directed his accolades to its high human and technological drama. He found Chicago's first home for the unit, the Merchandise Mart, then "the World's Largest Building," a fit site for this equally large-scale enterprise. Here, Reitman waxed, "The commodities of the mills, mines, factories and department stores are bought and sold without any merchandise to deliver, but their orders keep a thousand clerks busy. . . . On the sixth floor of this building is the WPA Statistical Division and the Headquarters for the survey of the Chicago Syphilis Control Program. Hundreds of clerks and dozens of rapid machines are tabulating all of the marvelous statistics and charts that have gone to make the Chicago 1937 Survey the most elaborate and far reaching syphilis survey every undertaken in the history of the World."[25]

The program's work had historical significance that would reflect to the credit of both the Chicago Board of Health and the U.S. Public Health Service. "Without the aid and competence of these cooperating groups," Reitman continued, "the Board of Health would have been unable to conduct any satisfactory survey, even in a limited degree. Whatever city in America or in the world undertakes a venereal survey will be obliged to follow the pattern laid down by Chicago in 1937."[26] He and Linck both demonstrated special sensitivity in addressing their remarks to a variety of groups, but the benefits to all groups were nonetheless real. Moreover, both men coupled an excitement with the scientific achievement of the project with an appreciation for the human dimension that the epidemiological study was to serve.

"A Bit Misleading"

One of Reitman's tasks as an inspector for the Social Security Administration was to examine and reflect upon the numbers generated by and relating to the Chicago Syphilis Control Project.[27] The tabulating effort fascinated him from the start, but he also advised caution in interpreting and using statistics. "The new headquarters of the Chicago Syphilis Survey on the second floor of the Merchandise Mart is chuck full of interesting material for any amateur mathematical syphilographer," he observed. "The charts and the figures provoke one to ask many questions and to challenge the authenticity of the data which one sees there. Dr Linck, and his entire staff, are so tolerant and so patient, they listen so thoroughly and explain so wisely that it is an inspiration to be with them. No matter how much or how little one is inclined to agree with the statistics one finds, one cannot question their integrity—they try to turn out accurate scientific statements."

He offered the familiar analogy with "two blind men who were describing an elephant—one had a hold on the trunk and the other the tail. They both described tails, but one was decidedly mistaken." In this case, Reitman described, the trunk of the statistical elephant was truth, but its tail was propaganda. "It will clarify our understanding of syphilis statistics very much," he went on, "if we can realize that most of the statistics are gathered for propaganda material and much of the data which is studied is interpreted BY THE NEWSPAPERS. For some unholy reason the newspapers and others have a mad desire to present syphilis in the loudest, largest, and most terrible aspects possible.... It behooves the statisticians of the Board of Health and the USPHS not to allow themselves to be entrapped by newspaper or syphilis propagandists who want a bogey-man."[28] Reitman directed his attack primarily at propagandizing newspapers, but his words of caution to his statistical colleagues were clear: Numbers are powerful tools, and such power should be wielded both consciously and conscientiously.

The early months of collecting and trying to interpret the statistics, even the ones being compiled by hand before the inauguration of the Central Tabulating Unit, probably raised as many questions as they provided answers. As the first compilations began to appear, Reitman tried to project what they would reveal.[29] When statistics released by the Chicago Board of Health in July 1937 showed fewer cases of syphilis reported than in June, he exulted, "There is every reason to believe that the peak of syphilis has been reached."[30] After four months of such reports, however, his jubilation gave way to skepticism. "We can not believe that this sudden drop in the number of positive darkfields is a true picture of the progress of the control of syphilis," he advised, noting that during September one of the most active clinicians was on vacation and that the chief of the clinicians had recently been transferred out of that position. "Whether there have been changes in the laboratory staff or not, we are unable to say," he concluded, but pointed out that the "rush of the clinic, the tremendous number of patients who must be seen daily is apt to make the average clinician neglectful."[31]

Of perennial concern to Reitman was the project's lack of control over and knowledge of all reporting sources. These concerns carried a special significance in the program's earlier months, when laboratory procedures and physicians' reporting forms were just beginning to be standardized. Besides the free city clinics and field stations, people also went to private clinics such as the Public Health Institute and the United Medical Service; other clinics associated with organizations or hospitals such as the Illinois Social Hygiene League, Cook County Hospital, and the Juvenile Detention Home; or their own private physicians. Although private doctors were required by law to report all new cases of syphilis, they did not always do so. When Reitman reported in mid-September that 311 cases of syphilis were gleaned from five thousand blood tests, he wondered how many private physicians were actually reporting all the cases they discovered.[32]

Although the Chicago Syphilis Control Program altered its strategy by focusing on their own clinics and the special teams they sent to schools, factories, and civic groups, there would never be full reporting of syphilis.

Clearly, exact numbers would never be available, but could an understanding of the epidemiology of syphilis be generalized from the information the project did gather?

Inadequacies of the facilities and personnel also bothered Reitman. As the numbers of people being tested began to increase significantly, he voiced concern that the resulting influx of numbers to be gathered and tabulated would overburden the statisticians and clinicians hired to do the work. Were there enough of them to do the careful, painstaking work required of them? For example, in a report entitled "The Pace That Kills," he described the work in the Board of Health Laboratory: "We sat in the laboratory and watched Director White checking up positives. His face was pale and drawn. He was engaged in serious business. . . . The group of workers in the serological room were working like automatons. The room was so crowded there was no place for a stool where these intelligent, sensitive, tired women might sit down."[33]

Moreover, because of the scope and novelty of the antisyphilis drive, its staff was constantly challenged to accommodate the statistical refinements that the medical and epidemiological needs of the program demanded, and the ways in which information was reported changed from time to time and further complicated the interpretation of results. For example, it wasn't always easy to keep clear the sources of information. The MSHC and City Hall Marriage Clinic both identified tests given in conjunction with an application for a marriage license and distinguished between tests that were "diagnostic" (to identify a new case of syphilis) or for "control" (repeated testing of a person already under treatment), however, private physicians often did not do this. Another example comes from the reporting forms, the first of which grouped all types of tertiary syphilis within the category "late," a designation that eventually proved inadequate and led to a subdivision of that stage. Communication between statisticians assigned to the Chicago Project and their supervisors, usually Lida Usilton, chief statistician of the Public Health Service, often dwelt on details of refining or adding categories to the various forms that the WPA had developed.

Reporting the results of tests was problematic for Chicago's newspaper writers as well. "The newspaper men haunt the President and the Secretary of the Board of Health for news," Reitman complained, aware of the frequent tension between timeliness and thoroughness. He went on, "And they want the latest news, the dope, they want to know how many people were examined and what percent were infected. . . . The results published in the TRIBUNE October 26th [1937] are a little bit misleading, and inaccurate. . . . It will be assumed by many that these tests are all poll tests, when the fact is that most of the positive tests are control tests of patients under treatment for syphilis. Many people will confuse these tests with the marriage tests. At the risk of annoying our bosses, we urge that more care be exercised in tabulating V.D. statistics." The importance to the public health officials of having accurate data was also clear to Reitman. "From the figures in the newspaper," he pointed out, "it would appear that 10 or 15% of everybody who has their blood tested has syphilis. This is far from the truth. So far as we have been able to estimate, less that 5% of the poll test is positive. And if we can keep our poll test separated from the rest of the tests, a good deal of valuable data will be secured."[34]

Reitman was differentiating between the "poll" test, people who responded to the questionnaire for whatever reason, and the people coming through the courts and were seen to be especially likely to have syphilis. The *Tribune* was citing percentages drawn from the combination of those two populations and calling this the rate of syphilis in Chicago, when in reality that number was unrealistically high. Moreover, because the original statistics didn't make this distinction, it would also be lost to the interpreters of the numbers.

Chicago officials did try to convey the complexity of the numbers to newspaper writers on some occasions, though. For example, "The percentage of positives is not considered representative of the general population," cautioned one article, qualifying its headline that 2,100 people had "react[ed]" to the blood test with the additional information that a second test was still required before reaching a final determination of the number of infections that had been discovered.[35] Another time, Bundesen said that recent reports had shown that one in eight persons being tested was found to have syphilis, one in five found infected with gonorrhea, but he quickly added that such ratios "must be qualified."[36] The interpretation of numbers was thus an exercise in both effective yet responsible reporting on the part of the newspapers and accurate yet useful epidemiology on the part of the public health officials.

How much, exactly, could those numbers reveal? Reitman, along with all of the interpreters of the endless numbers that flowed out of the accumulated tests and follow-up treatments, asked the question almost daily. Numbers may not lie, as the saying goes, but neither do they presume to interpret themselves. "Serological examinations, like unemployment," Reitman wryly observed, "becomes more complex and more difficult of solution the [more] it is investigated. The more books that are written on unemployment, the more statistics gathered, the more impossible seems the solution. The more blood tests, the more statistics, the more experts explaining things, the less clarity and the less surety."[37]

Besides the limitations of the numbers themselves, officials also faced the limits of scientific knowledge. In 1938, medical science was still learning about the pathology and manifestation of syphilis and related diseases. Moreover, the methods available to test for and treat these diseases were still subject to ongoing research and debate. In terms of the gathering and reporting of statistics, this uncertainty created special problems. For example, in a report titled "From Wassermann to Wilderness—From Kahn to Confusion," Reitman commented on the difficulty of categorizing disease status when medical science itself could not say with certainty when or if a case of syphilis had ceased to be infectious.[38] Whether the tests were identifying "old" or "new" cases of syphilis was also sometimes unclear, raising questions not only of subsequent treatment and tracing procedures but also of how to analyze the statistics.

Another problem with interpreting and reporting statistics had a direct impact on people's lives. For example, from the start, the project differentiated tests and treatments by race; statistics for syphilis and gonorrhea for "white" and "Negro" men and women were a source of comment in the earliest reports, with the incidence of both diseases continually higher among African Americans. How these numbers were related to syphilis's own relation to poverty was often not clearly discussed—nor was a comparison of the income levels of the two racial groups or the fact that most of the dragnet stations and city clinics were situated in poor, ethnic (largely African-American) neighborhoods. Public reports only of numbers of African Americans testing positive in comparison to whites, with no accompanying social or economic explanations, understandably triggered swift, angry response from Chicago's African-American community.[39]

Further underscoring the human element of statistics was the ever-present danger of error. Reitman discovered simple mistakes in some of the reports. For example, in August the U.S. Public Health Service credited Chicago with about a thousand more (more than twice as many) cases of syphilis than had been reported in the Chicago Board of Health's own report, the source of the Health Service's figures. "This is probably a clerical error," Reitman stated, mildly for him, but later added, with more typical irony, "Statistics have a habit of appearing in the most unexpected places and it might be well to ask the USPHS to correct its mistake."[40] In another report, he noted that the *Tribune* "confused insane patients with feeble-minded" ones in a story on the incidence of syphilis in state asylums. By incorrectly merging the two groups, the *Tribune* reported an incidence of syphilis in the "insane" that was about half its actual occurrence, because neurosyphilis was much more prevalent among the insane than the mentally retarded. "Never take figures from a newspaper," he concluded.[41] The error occurred even after an order had been issued by Health Commissioner Bundesen that "no figures be given out to the press, or anyone else, without being first carefully checked."[42] This edict was the upshot of another *Tribune* story in early November 1937 that reported that nineteen thousand cases of syphilis were under treatment in Chicago, when, in reality, a review of nineteen thousand patient *records* had revealed just over fourteen thousand patients in treatment.[43]

Although Reitman respected the fallibility of numbers and the people who handled them, he also respected the utility and power inherent in those numbers. As a part of his assignment as investigator, he pored over the statistics coming out of the Central Tabulating Unit daily, contemplating what they meant and how they might be used effectively but also accurately and fairly. It is notable that in his near-crusading enthusiasm to bring all people infected with syphilis under treatment he was so consistently anxious to avoid alarming misrepresentations of the presence and effects of the disease.

"More Direct Methods"

The problems inherent in interpreting data become magnified when that exercise is made public, adding concerns about public accountability to concerns about statistical reliability and validity. However cautious officials were in drawing conclusions from their new numbers, all of the information coming from the Chicago Syphilis Control Project to the people of Chicago through the *Tribune* suggested that the program was moving into another phase. In mid-July 1938, the *Tribune* reported that although the numbers of tests taken each month continued to rise, the number of new cases of infectious syphilis reported each month was declining. "Syphilis experts," the paper said, "take this as an indication that the rate of spread is declining." Similarly, in early August 1938, the paper announced "Syphilis War Shows Advance on Two Fronts" as the dragnet stations were beginning to be closed down. Although more cases than ever before were under treatment, fewer were being discovered in infectious stages. The reporter echoed the experts' opinion of the previous month, that these statistics showed that Chicago had succeeded in halting the spread of new disease.[44]

In his report in April of 1938, however, O. C. Wenger was not equating a decline in new cases with the halt of syphilis's spread. Instead, he read these numbers to indicate the limits of the usefulness of this part of the Chicago Syphilis Control Program. "From the evidence presented in this report," he concluded, "it can be said with complete assurance that the 'uncover and treat' method alone will not only prove inadequate, but become so complicated, so expensive, and so time-consuming that more direct methods must be found and used." Stating that typhoid, smallpox, diphtheria, and ophthalmia neonatorum (blindness in newborns caused by eye infection contracted during the birth process) had been conquered only when one was treated, voluntarily or not, Wenger warned that syphilis would be no exception. There was one problem: "Unfortunately we have no vaccine, no toxoid, for syphilis as we have for smallpox and diphtheria. We do have both mechanical and chemical prophylaxis, which, when properly used, prevent these diseases." He went on to push his recommendation, "For years medical men have known of these measures. The armies of the world have used them with remarkably good results. It is believed that the time is ripe and the American public in a receptive mood for the dissemination of such information as general knowledge and it has therefore been proposed to add to the Chicago slogan, 'Uncover and Treat,' the new word 'Prevent'—if you can;—if not—'Uncover and Treat.' Fires are prevented in this community, why not syphilis and gonorrhea?"[45]

The statistics being generated from the Chicago Syphilis Control Project also led Ben Reitman to see proof of the need for mechanical and chemical prophylaxis, although his conclusion might have been more predictable than Wenger's. "Figures, figures, figures!" he exploded at one point. "All lead to the definite statement that the amount of syphilis and gonorrhea in any community depends upon the attitude of the Health Authorities toward prophylaxis."[46]

Where Reitman had been told by members of the Chicago Board of Health to "soft-pedal" his push for prophylaxis almost a year earlier, Wenger now received a similar message from even higher levels within the program. A month after submitting his report, Wenger recommended to R. A. Vonderlehr that a special commission be appointed to "study the prophylaxis of the venereal diseases." Vonderlehr responded, "All of the experimental and practical work which has been done in this field indicates that the methods are not of particular value. . . . I should not be inclined to recommend to the Surgeon General that a committee be appointed unless some new preparation of known composition becomes available for prophylaxis."[47] But the surgeon general himself responded directly to Wenger's report. Parran wrote, "I have noticed your comment on the possibility of applying prophylactic procedures in Chicago. Until prophylactic measures known to be of greater efficacy than any known to science at the present time become available, I hope that you will hold this phase of the campaign in abeyance."[48] By at least April 1938, then, Ben Reitman was not alone in advocating the inclusion of prophylaxis to a program whose popular referendum—to test voluntarily all citizens of Chicago—had taken itself about as far as it could go.

The statistical work of the Chicago Syphilis Control Project, however, continued to serve the Public Health Service, both for the information it generated and as a model for other epidemiological statistical units, for some time to come. "The Chicago material provides the best statistical data so far available from mass blood testing," Parran wrote to Wenger as late as 1941.[49] That same year, Herman M. Soloway, chief of the Division of Venereal Disease for the Illinois Department of Public Health, boasted that "Illinois is one of the very few States in America to present facts and figures relating to the management and outcome of syphilitic pregnant women reported by and under the care of private physicians."[50] Indeed, the Chicago Syphilis Control Project brought a new rigor to epidemiological reporting. At the same time, it demonstrated only too well the human contexts of this science. The numbers the project generated emerged amid political squabbling, elicited elation or consternation depending upon how they were read, and failed to generate a consensus about what should be done next.

CHAPTER SEVEN

Privacy

Although by late 1938 statistics suggested that much of the syphilis being caught in Chicago's dragnet was old, noninfectious syphilis, the human circumstances surrounding the work of the Chicago Syphilis Control Program raised issues that numbers alone could not resolve. In its push for wider testing, concerns of privacy arose as the workplace and even schoolroom became centers for screening. The location of dragnet stations in lower socioeconomic, primarily ethnic neighborhoods, and the reporting of statistics by race raised questions about the relationship of disease to poverty and the dangers of privileging certain groups of people over others. Ongoing efforts to solicit the participation of private physicians and clinics led to questions about the economics of medical practice. Invoking fear of infection as a means of bringing citizens to the dragnet stations continued to contradict medical and journalistic desires to remove any stigmatizing onus from syphilis. And, finally, prevention continued to spark debate within the program, both in terms of its medical possibility and moral desirability.

Privacy, privilege, profit, paranoia, prophylaxis: to this alliteration of issues a sixth word could be added—*politics.* Certainly, goals are often formed and decisions directed by the people whose view determines how the world is seen and whose hands are on the purse strings. Taken together, these words encompass a range of attitudes about sexuality, morality, and social, racial, and economic privilege that underlay—at times intentionally and at times, athough no less seriously, unintentionally—the work of the Chicago Syphilis Control Program.

"The Duty of Every Good Citizen"

"It is the duty of every good citizen to take a blood test," the *Tribune* exhorted its readers.[1] Duty can be either required or merely expected. *Webster's Dictionary* defines *duty* as either "a moral or legal obligation." In its strictest legal sense, a duty is a tax, and, certainly, paying taxes, as a law, is a legislated "duty" of most citizens. In this sense, some of the duties of citizenship are behaviors that have been made enforceable. Other duties, sometimes more euphemistically referred to as the privileges of citizenship, do not carry the added weight of enforcement, such as voting or taking military service during times of nonconscription. "The duty of every good citizen," as directed by the *Tribune*, would imply this second, broader sense of duty, that of moral obligation. Indeed, the official rhetoric of the Chicago Syphilis Control Program—that citizens had before them a generous opportunity to avail themselves of the free tests—conveys this sense of a special privilege. Thus, the *Tribune*'s use of the term *good citizen* in conjunction with the word *duty* implies not only this broader, unenforced sense but its moral weight as well.

Public health offices or departments, though, also possess the ability to enforce duty, to police compliance with law. In fact, as O. C. Wenger pointed out in his first annual report, such measures had already been taken in regard to a number of infectious or communicable diseases. The difference with syphilis, he acknowledged, was that there was no single-dose preventive measure as there was for smallpox or diphtheria. A mandatory, universal test, in the case of syphilis, was not the same thing as a one-time, mandatory, universal vaccination or inoculation, because a sexually active person would have repeated opportunities to become infected. In the light of such uncertainties, then, what could or should public health, as an arm of the government, require of its constituency?

These questions will come up again around issues of prevention, but in terms of mass testing and treatment, the U.S. Public Health Service looked not to its own policing powers but to the powers possessed of businesses and other fundamentally self-governing organizations such as school systems and universities to "legislate" their own "laws"—create their own policies—regarding syphilis. Testing the staff of the *Chicago Tribune* is a perfect illustration. To begin, current employees were not required to take the test. The *Tribune* staff itself voted that all of its present employees should be tested by May 1, 1938, by either the newspaper's own physician at company expense or by a private physician of the employee's choice and at the employee's expense. There*after*, all new employees would be required to take the Kahn test for syphilis at the time they were hired. An important qualification to the new requirement existed, however: "The result of these tests will not determine employment or rejection of the individual," the news story pointed out, "since THE TRIBUNE recognizes that although such a test may show an infection, proper treatment will insure protection of other persons from infection. The infected employé will not, however, be admitted to The Tribune's sick benefit plan until treatment indicates his cure."[2] The newspaper's directors assured employees that they would not fire or refuse to hire anyone who was discovered to be infected, and they emphasized the benefits to the workplace of finding and treating syphilis.

Testing the *Tribune*'s staff was more than a publicity stunt. It was part of a special new program within the Public Health Service: Syphilis Control in Industry was created in November 1937 and run by specialists in industrial hygiene. As promoted in the surgeon general's annual report for 1938, manufacturers would be asked to test both current and new employees, with the understanding that employees' health status would not affect their employability. Workers with positive test results would be advised by the plant physician, who would send them to workers' private physicians or to one of the low- or no-cost clinics. Workers would need to furnish proof of ongoing or successfully completed treatment to continue employment.[3] The *Tribune*'s assurance to its own employees that anyone discovered to have syphilis would not be fired reflects these guidelines. The assurance of continued employment, however, was not necessarily required by law, a fact of which the Public Health Service and Chicago Syphilis Control Program were well aware. Newspaper stories about testing in the workplace repeatedly insisted that discrimination against infected employees or applicants was reprehensible and mean-spirited.

Syphilis Control in Industry received constant news coverage in March and April of 1938, as the *Tribune* printed, in quick succession, stories of businesses signing up for employee testing, among them Sears-Roebuck, Montgomery Ward, Bowman Dairy, Charles A. Stevens (a women's clothing company), and Carson Pirie Scott (a major Chicago department store). Local unions that supported the program included carpenters and joiners, window washers, metal polishers, bakers and confectioners, porters, barbers, boot and shoe workers, and belt, pocketbook and luggage workers.[4] In the following months, the *Tribune* continued its litany, as Dry Zero (a dry ice plant), Caterpillar, the Illinois Manufacturers Association, the Chicago Laundry Owners Association, and Chicago Restaurant Association pledged support or described their testing programs.[5]

The roll call was impressive, but, considering that there were 2,800 factories in Illinois at that time, such a response was quite moderate. "Chicago is notoriously difficult," Wenger reported six months after Syphilis Control in Industry was created. Indeed, he was constantly being called upon "to iron out the little difficulties that are always cropping up, either from labor or from the employer."[6]

Although Syphilis Control in Industry was officially declared to have "been received with enthusiasm by a number of industrial organizations," people involved in Chicago's program expressed more apprehension when they communicated informally.[7] For example, even before the creation of Syphilis Control in Industry, R. A. Vonderlehr had written to David C. Elliott, one of the unit's leaders, that he would provide testing for industry only if companies would agree not to fire infected employees, a demand, it turned out, that was not so easily met.[8]

Another officer of Syphilis Control in Industry, Albert E. Russell, reported a frustrating visit to Chicago in March 1938, where he found local industrial leaders aggravatingly reluctant to embrace the government's program. He wrote to Surgeon General Thomas Parran, "Doctor Wenger was a bit disconcerted because I had not had better results in getting the companies to consent for blood testing. He feels that the year of publicity which the Chicago Tribune has given should have convinced the industrialists of the value of blood testing."[9] There were two reasons for this hesitation, one political and the other economic. On the political front, Russell cited as typical his conversation with high officials in Donnelly Printing (which printed such magazines as *Time*, *Life*, and *Fortune*), who said quite straightforwardly that they were "interested" in syphilis but didn't like either Chicago's city administration in general or Herman Bundesen in particular.[10]

Certainly, Chicago's traditionally reform, labor-oriented administrations were often viewed askance by the city's industrialists, but companies' managers also gave another reason for hesitancy: the inherent contradiction they saw in being asked to put on their payroll people who were felt to be economic risks. Wenger summarized the conflict nicely:

> The majority [of industrialists] states that they are interested in syphilis only because they are convinced that it is an industrial hazard and

> they are perfectly willing to be responsible for the diagnosis and treatment of those cases now within their ranks. They cannot understand, however, why they should be expected to be responsible for the potential hazard and liability which would be incurred by the employment of additional infected employees. Certainly we are not consistent when we insist syphilis is a serious industrial hazard and then recommend that industry employ infected persons with[out] prejudice.[11]

Wenger was referring in his last sentence to the kind of publicity that reported that twenty-one million days of work were lost annually to syphilis—in the same article in which Schmidt was "caution[ing] employers against discharge of the syphilitic employé."[12] Employers should not "elect to test" employees only to "discharge those found to be infected," another article pressed, and in another story Dr. Wenger himself argued, "Protection and not penalization [is] a function of good business."[13] Thus, Wenger was painfully aware, while program officials insisted on the long-term economic advantages of testing current and potential employees, industry often saw only the immediate cost of infected workers.

On the other side of the coin, and no less surprisingly, labor was suspicious of employers' motives for any testing. "Labor, in most instances," Wenger reported, "objects to the taking of blood specimens by the Company physician. The feeling is that the physician, paid by the Company, has only the Company's interest at heart. The Unions are apprehensive that positive blood test reports would result in discrimination against the persons found so infected, would certainly prevent their advancement and possibly mean their divorce from the payrolls."

In general, however, Wenger found that unions, although still "more or less suspicious," were willing to consider testing "and will co-operate if we can work out a satisfactory plan for the protection of their members."[14] Indeed, in March 1938, Chicago's labor groups backed the program, with 150 delegates from local unions pledging to play "their part in the campaign." John Fitzpatrick, president of the Chicago Federation of Labor, "urged immediate cooperation with employers and the city board of health in having tests made before employers demanded such action." He made this recommendation, he said, so that "those who happen to be suffering from the disease may get proper treatment." Joseph D. Keenan, the federation's secretary, was also quoted: "There is a growing realization among the unions of a need for education on the prevention and control of syphilis."[15]

In the final analysis Wenger felt that management was "the biggest obstacle we have to meet in Chicago." Such a statement indicates that even by April 1938, Wenger saw the volunteer effort, including the dragnet stations, as making only a minor contribution to the goals of the Chicago Syphilis Control Program. Indeed, the importance of creating mandatory testing and treatment through work programs loomed largest in the success or failure of the program. "Up to the present we have hedged the question," he concluded, "but sooner or later some compromise must be reached; otherwise we cannot expect the full support of the industrial groups."[16]

The controversy posed by syphilis testing in business was brought home to the Chicago Syphilis Control Program in November 1938, when the *Chicago Daily News* ran a story under the headline, "Laundry Workers Head the List." Ostensibly only reporting the latest numbers released by the Chicago Syphilis Control Project, the newspaper noted that laundry personnel had the highest percentage of infection among the business groups tested to date. The complaints that the article generated reached all the way to R. A. Vonderlehr in Washington, who wrote to Wenger back in Chicago that "the laundries were deluged with calls from customers asking if any syphilitic workers were employed" and asked him to "make tactful inquiry" into the situation and provide more information.[17] The Chicago Laundry Association, angry over such publicity, promptly canceled its syphilis control program, which had begun only three months before.[18] In this instance, both labor and management felt the unpleasant consequences of thoughtless publicity around syphilis control. Obviously, and however wrongheadedly, the syphilis of many blue-collar workers mattered to the general public, and thus mattered in terms of profits and continued employment.

This story had another aspect. Lawrence Linck, director of the statistical project and the person responsible for the statistics released to the *Daily News*, left his position a few months after Vonderlehr asked Wenger to investigate "tactfully." Upon the recommendation of senior statistician and director of the project in the Washington office, Lida Usilton, Linck was replaced by Dr. Bertha Shafer, director of the Illinois Social Hygiene League and a widely respected researcher of and educator about venereal disease. The exact circumstances of Linck's departure are not stated, but Usilton commented to David Elliott at one point that "the best thing for [Linck] to do was to seek employment in some field outside syphilis."[19] Although it is not clear that Usilton's comment referred to Linck's statistical acumen, managerial abilities, or sensitivity to the public discourse around venereal disease, her comment, added to Vonderlehr's apparent displeasure over the laundries' insult, suggests that the federal overseers of the program were aware that the work of a statistician in a public health program far transcends collecting and tabulating data.

The insurance industry presented syphilis control with another, similar obstacle to the exercise of "good," responsible citizenship. The debate was alluded to in the *Tribune*'s story about testing its employees. A sentence toward the end of the story remarked that new employees discovered to be infected with syphilis would not be admitted to the *Tribune*'s health plan until their treatment was complete.[20] The policy reflected an ongoing controversy. Two months earlier, the *Tribune* had quoted a statement of Dr. Louis Schmidt, who was reported to have said that people infected with syphilis should not be insured, and that if companies denied coverage to anyone with a venereal infection, insurance premiums for all policyholders would consequently be reduced. "The only people who would object to taking the tests are those suffering from syphilis and those people shouldn't get life insurance," Schmidt was quoted as saying. "We hope to enact a law requiring all insurance companies to include the syphilis tests in all medical examinations before the companies will be licensed to do business in this state." Ben Reitman responded with dismay and disbelief: "Christ help us! Let us hope that the Secretary of the Chicago Board of Health is incorrectly quoted. Already it is thought by industry that syphilitics shouldn't get jobs and now if we deny them life insurance, we will make the lot of the syphilitic a little harder."[21]

A week later, he commented with equal anger to an editorial in the *Tribune* that called for mandatory testing of anyone applying for a life insurance policy of more than four to five thousand dollars because "universal tests would be a measure of justice to all insured."[22] Schmidt was again cited, but for his support of mandatory testing only. Reitman realized that Schmidt *had* been misquoted about not insuring people with syphilis and regretfully concluded that Schmidt had allowed himself to become a pawn of the commercial world.[23] Subsequent newspaper stories were modified to state only that people infected with syphilis should pay higher rates, but Reitman still objected. On February 13, 1938, the *Tribune* quoted Usilton, "If a syphilitic is a poorer risk than an uninfected person, he should not be able to buy insurance as cheaply." Reitman accused Usilton of being "the apparent mouthpiece for the Chicago Board of Health" and criticized her for permitting herself "to be quoted as having made a statement that is detrimental to the syphilitic." He insisted, "IF PROGRESS IS TO BE MADE IN THE CONTROL OF SYPHILIS THE COOPERATION OF THE PATIENT MUST BE SECURED."[24]

One of Reitman's greatest fears in this matter of insurance was that such policies would drive infected men and women underground rather than encourage them to seek information about their health. In a similar vein, he urged that employees already carrying insurance be clearly informed that not only would they receive compensation when a work injury aggravated their venereal disease but also that they would not be held liable for any accidents caused by a syphilis-related impairment. Such assurances were strong arguments for being tested, Reitman held.[25] Finally, he suggested that involving insurance companies in the expense of venereal disease in a quite different way might speed the accomplishment of two goals near to his heart: universalized health care and public education on the mechanical and chemical prevention of venereal disease. "Whenever it is possible," Reitman advised, "the burden of the VD patient should be shifted from the taxpayer to the Insurance Co. The idea of the german 'kraken casse' insurance for all people is growing in America. Perhaps some of these insurance companies will help prop[o]gate venereal prophylaxis."[26]

Some officials of the federal government, in fact, were equally annoyed at the power some insurance companies seemed to be plying in industry. For example, in late 1937 David Elliott discovered that insurance companies were influencing the policies of some businesses regarding the hiring and retention of employees with syphilis. Whether they were doing this through general coverage of a plant, workman's compensation programs, or other employee health benefits is not clear, but Elliott's response was sharp and his recommendation clearly threatening. "It is . . . believed that if further evidence of discrimination by industries, against employees found to be infected with syphilis, is developed and presented to [the state insurance department]," he wrote to Parran, "that this practice can be eliminated by refusing to grant to such insurance companies the privilege of writing coverage in this State."[27]

Both Reitman and Elliott were exploring ways to make the protection of individuals being tested for syphilis enforceable through legislation directed at business rather than individuals. Such laws, both men reasoned, would make citizens' moral duty to be tested for syphilis more attractive if employers had a legal duty to treat and not expose or punish infected people. The fact that Reitman and his bosses both recognized that it was the worker who was at most risk economically by such an exercise of good citizenship underscores how unwilling many people still were to reason generously about syphilis.

"Essentially a Youth Problem"

Syphilis Control in Industry argued for mass testing in terms of the economic benefits to business, benefits great enough to justify protecting the privacy of and providing health care for employees. In extending syphilis testing to children and young adults, arguments were framed differently. For young adults, the virtue of voluntarism was a major theme; with children, protestations of public concern for protecting them were most often heard. Although privacy was never a major issue in the testing of either of these groups, it nevertheless maintained a shadowy presence.

"Syphilis is essentially a youth problem, and probably one-half of the syphilis victims acquire the disease innocently," Bertha Shafer told a group of students at the Central YMCA College at a special assembly during the summer of the program's inauguration.[28] Education was even more important to the program's work with youth than with other age groups, because from the start it was expected that the rates of syphilis in schoolchildren would be low. The Illinois Social Hygiene League, as part of the American Social Hygiene Association, had long supported sex education, so it staunchly advocated educating youth as an essential first step in eradicating syphilis.[29]

School programs for syphilis testing appeared in colleges and secondary schools. In the former, students themselves, usually through the government's American Youth Congress or colleges' own American Student Health organizations, often led the drive to include syphilis testing as either a part of their school's available health services or even an entrance requirement.[30] Concern about use of test results was voiced, as it had been in industry, and, similarly, the Public Health Service recommended that such information should not be used as a basis for refusing admission.[31] The advised protocol, adopted by the YMCA College and the University of Chicago during the school year that ended in 1939, was to require a serologic test for all new students. Those testing positive would be put into treatment through the Student Health Service.[32]

The University of Chicago was noteworthy not only for installing this requirement but also for requiring it even for students enrolled in its summer programs. Wenger felt that the policy was particularly important "because these students are an adult group of teachers, and others, who take the idea back to their home communities."[33] When Northwestern University students drafted their own plan to educate themselves about syphilis, they were praised for their initiative. Their student congress "believe[d] itself to be the first student body in any American university to promote venereal disease education."[34] Calling it a "civic responsibility of the students to help carry forward the campaign of eradication . . . as a matter of self-protection," the *Tribune*'s health columnist, Dr. Irving Cutter, himself on the faculty of Northwestern University's Medical School, addressed the students at a special meeting, after which Bertha Shafer demonstrated the administration of the Wassermann examination. Three hundred fifty students volunteered on the spot to be tested.[35]

At that time Northwestern's policy did not permit entrance of students known to be infected with syphilis, making those who volunteered appear even more courageous. Other schools, such as the University of Chicago, University of Iowa, and Dartmouth University, tested all new students but only placed infected students under treatment. The University of Illinois, in Urbana, tested only those students who exhibited "suggestive symptoms." Policies varied widely among schools, but most schools favored some form of sex education over compulsory testing. Although, the *Tribune* reported, most university presidents or deans believed that infected students should be kept in school, most of them also commented that they did not see syphilis as much of a campus problem.[36] Schmidt responded immediately to this seeming dismissal, and the *Tribune* expressed his irritation in bold letters: "Attacks Stand of Educators on Syphilis Menace." Even though the incidence of two cases per thousand was below the suspected rate in the general population, he pointed out, *all* syphilis was a problem. Besides, unknown cases were a bigger menace than known cases, and the lack of a testing program in universities could contribute to larger, undetected rates of infection. Purdue University, Schmidt added, had been regularly examining students even before the current campaign began.[37]

A curious, but not particularly surprising, variation on the theme of testing and education in a collegiate setting occurred in downstate Illinois and was prominently covered by the *Tribune* in the fall of the program's first year. "U. of Illinois Joins Student Paper's Anti-Syphilis Fight," Chicagoans read, learning of a recent exposé in the school's student newspaper. An article based on recently released statistics from the State Board of Health reported that 70 percent of the customers of the town's houses of prostitution were university students. The resulting antisyphilis drive on campus was reminiscent of the antiprostitution drives of earlier decades. University officials responded to (at least some) students' indignation over the matter by stating that any student apprehended in one of fourteen local houses of prostitution would be expelled immediately. Three days later, the

Tribune announced that three "dives" had been closed but twelve remained open as "Police Fail to Make Raids." Two days after that, two students were expelled from the university in a "Cleanup of Vice"; action on three others caught in the raid, however, was deferred.[38]

After this brief flurry of police activity, always printed in the first few pages of the *Tribune*, no more such stories appeared in the Chicago paper, but it is probably fair to assume that most houses of prostitution remained open, and that University of Illinois students, whatever percentage of the business they constituted, were not deterred for long. Such an approach to syphilis control should not be surprising. In addition to historical precedence, the American Social Hygiene Association was the main body working with college students, and it had come to antisyphilis work through its efforts against prostitution. The role of student leaders as moral leaders—and syphilis as inextricable from social mores—would have been reinforced by this connection as well as by Parran's exhortations to clean living in America's young adults.

Syphilis test as object lesson, moreover, was a notion that the program's officials repeated often in work in secondary schools. "Once the child learns to associate the prick of the needle with syphilis, and is made to realize that two years' treatment with such needles is necessary to cure the disease, I believe that exposures in later life will be further apart," averred Dr. Wenger. "In other words, I believe that there is more Social Hygiene and sex education remembered from the prick of a needle than by many lectures on the same subject."[39] In November 1937, the Board of Health began distributing cards that invited all students beyond the sixth grade to be tested. "Authorization for Free Blood Test," the card read, and students were directed to ask their parents for signed permission. A serological team of physician, nurse, and clerk would then visit schools to perform the test. By April 1938, four of Chicago's forty-seven high schools had hosted the team, with Montgomery Ward High School being the first. On November 2, 1938, health teams tested a thousand teenagers at that school, to high public praise.[40]

The school program was unique to Chicago. It was, officials claimed, "the first attempt to collect serological tests from large groups of school children and it points the way to uncovering hidden congenital syphilis in this group."[41] Schmidt was quick to point out, however, "that the school group tests are not intended to quarantine or embarrass families, but are protection for the welfare and future of the children."[42] Response in the first two schools was greater than in the second two, a fact that Wenger qualified with the observation, "It must also be taken into consideration that the first two schools were surveyed in November while the publicity campaign was in full swing and public interest was high. . . . The lack of response [in the schools visited after January] is partially due to lapsed interest in the subject."[43]

Wenger's comment is important for two reasons. First, he was being deliberately careful not to make distinctions in the test results based on race, conscious that enrollment in the first two schools was predominately African American. Second, his comment revealed that a "lapse" of even two months after the promotion of an activity could sharply affect participation. As with the drop in public enthusiasm for the dragnet stations, the city school testing program relied heavily on public support, which had a relatively short attention span. Nonetheless, officials felt that the results to be gained—both medically and educationally—were worth the effort. Testing continued into the spring; and in May the *Tribune* reported that testing teams had gone into three more Chicago high schools, where they had received parents' permission to test 1,100 students at Englewood, 400 at Hyde Park, and 400 at Lane Tech.[44] When the new school year began in the fall of 1938, the *Tribune* announced that schools were being urged to resume, energetically, the voluntary testing program of the previous winter.[45]

Assurances of confidentiality and insistence on the safety and necessity of keeping infected children in school did not occur with the same frequency with which these themes were sounded in discussions of testing adult populations in work or study settings, however, and there were no reports of students being barred from school if they were discovered to have syphilis. One reason for this could be that it was presumed that most of the syphilis being detected among schoolchildren was congenital syphilis, which was not in an infectious stage. The frequent argument that the test, in fact, was a boon to children because it could detect syphilis before the disease had begun its most deadly work probably led testing in this setting to be seen in a more beneficent and less threatening light than it was often perceived in other settings.

The added benefit of the test, as Wenger observed, was as a warning to all students against what might happen when they became sexually active.[46] Other health leaders, however, were making more radical proposals in the area of sex education. Dr. A. J. Levy, an epidemiologist in the state Health Department, addressing the Illinois State Medical Society, recommended that young people be taught how to use mechanical and chemical means of preventing syphilis. "When prophylaxis is practiced intelligently among our youth," Levy said, "it will considerably reduce venereal diseases. Every youth over 17 should be taught to apply prophylactic measures."[47] This kind of sex education, however, was not what was implied when Wenger wrote about the lesson to be learned from the prick of the testing needle. Where Levy would teach individual responsibility and choice, Wenger's lessons implied fear and obedience to impersonal authority.

The difference between these two types of education, however, had come up in public debate six months earlier, and the Chicago Syphilis Control Program had apparently conceded to the city's educational leaders. About the program's request that "Schools . . . Give Pupils Syphilis Facts," the *Tribune* reported, "School authorities in reply explained that tax-supported education must not get too far ahead of public opinion for fear of an unfavorable reaction. According to educators, parents in general have not shown a readiness to have their children receive sex instruction and are even less inclined toward teaching the nature and dangers of syphilis."[48]

Thus, what was *done* to students, in terms of a syphilis test, was less inflammatory to many people than what children might be *told* about sex and syphilis. Sex education also raises issues of privacy; most of the debate turned around what parents wanted their children to know, with sex education often seen as an intrusion on the privacy of family. The argument of family was used in other ways as well. In the case of younger schoolchildren, the notion of duty was presented to *parents* as a familial duty to protect their innocent children—as Schmidt had put it, "protection for the welfare and future of the children."[49] These children's participation in the Chicago Syphilis Control Program was usually described in emotional terms, as they were hailed as contributing to a better future for themselves and their own future daughters and sons.

To the Children's Aid

Demands from college students that their schools create policies to require testing new students were similar to the labor unions' demands for testing and treatment programs, but the role of younger schoolchildren in the debates around required testing was more passive because they generally received the treatment their parents chose for them. In the schoolchildren's case, privacy—and the element of choice that adheres to that notion—remained primarily with adults, the children's parents. In the case of children under the age of majority, their very youth—and supposed innocence—became the theme: unsuspecting children who had contracted the disease through no act of their own. Children outside of traditional family situations, however, found themselves placed in even more passive—and more vulnerable—positions. Although the following two situations are almost polar opposites, they raise the problem of honoring or maintaining the privacy of some of the children who came in contact with the Chicago Syphilis Control Program.

In July 1937, around the same time that the second survey of physicians treating patients with syphilis had begun, the Chicago Board of Health tested all the children being cared for in Chicago through the Illinois Children's Home and Aid Society. This testing marked another "first" in the program's initiatives: the first time any state had tested its underage wards.[50] An editorial in the *Tribune* applauded the undertaking, suggesting even broader application to other wards of state institutions if the blanket testing turned up more cases than were found under the current system, which required testing children only in instances of suspected syphilis.[51] By the beginning of 1938, testing at selected sites was revealing less than 1 percent of the children to be infected, yet, although the *Tribune* did not report the former percentages based on the old testing practices, the state was preparing to expand the tests, which had initially been performed only on its wards in Cook County, to children across the state.[52] At the same time, officials of a private organization, the Frances Juvenile Home, told the *Tribune* that they had been testing and treating for syphilis the girls they had taken in since the home was founded in 1909. Twenty-four girls were currently being treated and, reported Doctors Yarros and Shafer, taught about syphilis in the hope of preventing future infection.[53] In both cases, minors subject to the support of charitable institutions became immediately subject to decisions about their health and welfare, decisions one step removed from individual parents' decisions about the testing of their children.

The second instance of special vulnerability shows a similar lack of self-determination, but this time in a noninstitutionalized, independent minor. In his first annual report, O. C. Wenger presented the following complaint about the dragnet stations: "The greatest obstacle that we have encountered in this method of taking specimens is that under Illinois law no person under 21 years of age may be vaccinated, inoculated, or operated upon without the consent of parent or guardian. Only imagine the embarrassment of the physician who had to inform a strong, young woman, mother of two children, just short of her twenty-first birthday, that she must bring written consent from parent or guardian before her specimen could be taken!"[54] In such cases, existing law could actually prevent a "good" citizen from performing her "moral" duty, could actually embarrass (or physically endanger) her because her lack of privacy was denied by law.

Honoring the privacy of individuals in knowing the results of their syphilis tests or even deciding whether to have the test was often inextricable from the social and economic circumstances of their lives. Required testing, although perhaps benefiting people's physical health, could also subject them to greater public scrutiny, innuendo, or censure than would fall upon people who could take knowledge about their health into their own hands and keep that knowledge private. After all, as the officers and many other people involved in the work of the Chicago Syphilis Control Program realized only too clearly, even when the government thought it could—or should—legislate policy (or influence other organizations to legislate their own), no organization could legislate fair-minded or compassionate public opinion.

Oliver Clarence Wegner, special consultant, U.S. Public Health Service, to the Chicago Syphilis Control Program (circa 1930). (National Library of Medicine)

Thomas Parran (right), U.S. surgeon general, speaking to Raymond A. Vonderlehr, head, Division of Venereal Diseases, U.S. Public Health Services (circa 1940). (National Library of Medicine)

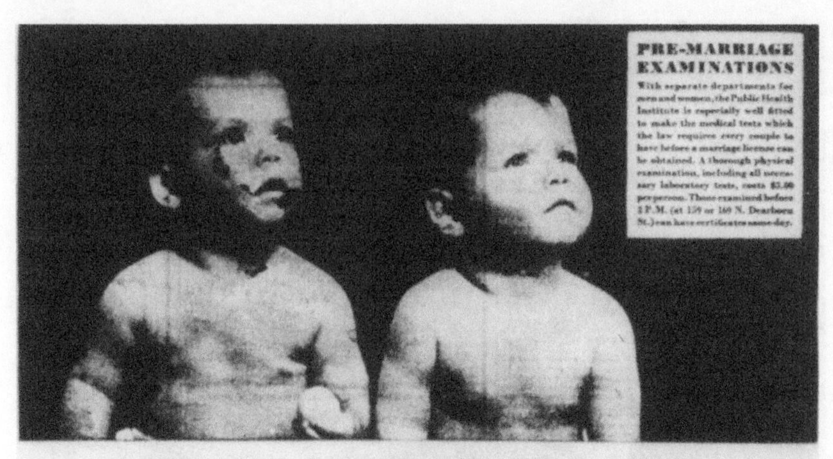

THEY NEEDN'T DIE!

Shield *your* baby from the deadly effects of unsuspected syphilis by being examined early in pregnancy by your physician or this clinic!

MEDICAL authorities have found that *nine out of every ten* babies of mothers with untreated syphilis are born dead or diseased!

And they have also found that of the babies who are born alive but diseased, *only about one-fourth* live beyond the first year unless given proper treatment!

Doesn't this suggest the wisdom of being examined for syphilis *early* in your pregnancy? For if infection is discovered before the fifth month, treatment can be started which will give you *ten chances in eleven* of bearing a healthy child!

Remember, it is entirely possible for you to be infected and not even suspect it. For syphilis can be innocently contracted, from drinking cups, towels, eating utensils, etc., and exist for months or years without your knowledge.

Mothers can also suffer!

Aside from its deadly effects on babies, untreated syphilis can be the cause of serious trouble in mothers. It may, for example, lead to heart disease, blindness, deafness, paralysis, or even insanity! Untreated gonorrhea is likewise a source of grave danger, as it can easily blind an infant's eyes at birth unless certain precautions are taken. Also, it frequently condemns women to mutilating pelvic operations that make it impossible for them *ever* to have children!

For these reasons, why not make up your mind to be examined for syphilis and gonorrhea before another week has passed? Such an examination can be obtained from your family doctor, or you can get it confidentially and inexpensively in the Women's Department of the Public Health Institute.

The Public Health Institute is the largest low-cost venereal disease clinic in the entire world. Since 1920, it has given more than 5 million examinations and treatments for syphilis and gonorrhea! Thus you are sure of getting medical attention of *proved* dependability.

No one is turned away!

Because the Public Health Institute was organized by public-spirited citizens as a corporation "NOT-FOR-PROFIT" under the laws of the State of Illinois, it is able to provide expert venereal disease treatment at charges to fit the lowest income. The cost of a thorough examination for syphilis and gonorrhea, including blood tests, urinalysis and gonorrheal smears, is surprisingly moderate, while individual treatments are available at fees so small that any employed person can readily pay them!

Nor does the Public Health Institute stop there in its efforts to make dependable venereal disease attention available to the entire community. Even persons who are temporarily short of money can make arrangements to be cared for! Thus there is no reason why any responsible venereal disease sufferer should be deprived of treatment.

PUBLIC HEALTH INSTITUTE

Organized by Public-Spirited Citizens as a Corporation "NOT-FOR-PROFIT" Under the Laws of the State of Illinois

DEPARTMENT FOR MEN • 159 NORTH DEARBORN STREET, BETWEEN RANDOLPH AND LAKE

Department for Women and Children Only
160 North Dearborn Street, North of Randolph

South Side Department for Men and Women
129 East 31st Street, West of Indiana Avenue

ALL DEPARTMENTS OPEN FROM 10 A.M. TO 8 P.M. DAILY, EXCEPT SUNDAY. TELEPHONE CENTRAL [illegible]

The Public Health Institute advertises its services during the months of the Chicago Syphilis Control Program.

DON'T LOSE OUT

on good times this summer because of a neglected venereal infection!

Gonorrhea or syphilis needn't bar you from sports if you'll procure treatment from your family doctor or this reliable clinic—at once!

SUMMER is a season of happy days—days of swimming, biking, baseball, golf and other outdoor activities. But they hold little joy for sufferers from venereal diseases, as these infections make it extremely dangerous for their victims to take part in vigorous sports unless they are being properly treated.

Thus all men and women who have gonorrhea or syphilis should hasten to obtain dependable medical attention. For not only will they be barred from good times until they get under treatment, but *if they continue to neglect their infections they may become permanently disabled!*

Neglected gonorrhea, for example, may extend to the prostate, bladder or even kidneys, resulting in serious disorders. Or the germs may penetrate to the joints, causing crippling rheumatism. It is the foremost cause of sterility, and is responsible for most of the mutilating pelvic operations performed on women.

Neglected syphilis is still more dangerous, for it may lead to insanity, blindness, deafness, locomotor ataxia, apoplexy, paralysis, ulcers of the skin or bone destruction! Neglected syphilis is also accountable for more stillborn and defective babies than any other cause.

It's risky to wait a day!

Hence it's extremely reckless to neglect a venereal disease, or to try to cure it yourself. For the sooner a competent physician or clinic is consulted, the better are your chances of recovery. *Thus every month, week or even day you save has an important bearing on the final outcome of your case!*

Especially is this true of syphilis, for unless treatment is begun in its earlier stages it may not be possible to do any more than stop further progress of the disease. So, if you know or suspect that you have a venereal infection, don't continue to neglect it or aggravate it by self treatment but immediately consult your family doctor or the Public Health Institute!

The friendly, sympathetic men and women physicians of the Public Health Institute specialize in the examination and treatment of venereal diseases. To date they have given over 3,000,000 examinations and treatments to over 250,000 individual patients! Each patient is promptly attended in a private room, and for the sake of even greater privacy is known by a number instead of by name.

Treatments as low as $1.00

Organized by public-spirited citizens as a corporation "NOT-FOR-PROFIT" under the laws of the State of Illinois, the Public Health Institute offers low cost venereal disease treatment to those unable to pay higher fees. A complete examination for gonorrhea and syphilis, including urinalysis, smears and Wassermann blood test, costs only $5.00, while some treatments for syphilis cost as little as $1.00 and some of those for gonorrhea still less!

Thus it's foolish to allow a venereal infection to keep you from having good times this summer and perhaps disable you permanently. For not only can the very best of medical attention be obtained confidentially and at nominal cost, but even persons who are temporarily pressed for funds can arrange to be treated!

NOTE: *The Public Health Institute has no connection with any other Chicago health or medical institute, and should not be confused with those operated for personal gain!*

PUBLIC HEALTH INSTITUTE

Organized by Public Spirited Citizens as a Corporation **"NOT-FOR-PROFIT"** Under the Laws of the State of Illinois

DEPARTMENT FOR MEN • **159 NORTH DEARBORN STREET, BETWEEN RANDOLPH AND LAKE**

Department for Women and Children Only • ***South Side Department for Men and Women***
169 North Dearborn Street, North of Randolph • **129 East 31st Street, West of Indiana Avenue**

ALL DEPARTMENTS OPEN FROM 10 A.M. TO 9 P.M. DAILY, EXCEPT SUNDAY. TELEPHONE CENTRAL 1983

CHAPTER EIGHT

Privilege

Syphilis testing itself—whether voluntary or not—was an intrusion into people's privacy. It could change the course of lives in ways that no promises of confidentiality could prevent. Officials of the Chicago Syphilis Control Program were well aware that the consequences of being found to have the disease went far beyond becoming eligible for free drugs. They repeatedly insisted that no onus should attach to any individual infected with syphilis and pressured employers, insurance companies, and schools not to turn away infected men and women. They were less careful, however, in considering the consequences of syphilis to some of the ethnic, social, and racial groups they served.

Where privacy usually applies to individual situations, the notion of privilege involves the consequences of actions or attitudes on people by virtue of the various groups or classes by which they might be identified (e.g., African American, female, homosexual). The work of the Chicago Syphilis Control Program frequently raised issues that would now be defined as those of privilege. Some were recognized as such at the time, whereas others were either unrecognized or simply ignored. Identified as privilege or not, though the program's actual or potential power over unique groups or classes of people was often apparent to those on both sides of that differential.

"Approximately 10 Percent of Whom Are Colored"

On July 14, 1939, R. A. Vonderlehr, from his office in Washington, D.C., wrote to O. C. Wenger in Chicago about the possibility of conducting a study of untreated cardiovascular syphilis in Chicago. He explicitly likened the proposed study to one currently in progress through the Public Health Service's Cooperative Clinical Study with the Tuskegee Institute in Alabama.[1] Given the openness with which the Tuskegee Study was conducted and documented, the lack of records for such a study in Chicago offers strong indication that Vonderlehr's suggestion was not put into action. That the suggestion was even offered, however, is important, in terms of both the historical connections between the Chicago Syphilis Control Program and the Tuskegee Study and what this connection signifies for the role of race in Chicago's program.

In *Bad Blood: The Tuskegee Syphilis Experiment*, James H. Jones presents the tragedy of this study of syphilis in nearly four hundred African-American men in poor, rural Alabama; their syphilis was deliberately left untreated for the entire course of their lives, even after penicillin became an easy and standard treatment for the disease. As the Chicago Syphilis Control Program was beginning, Surgeon General Thomas Parran's annual reports to Congress were requesting money to perform autopsies on the first men of this cohort to die.[2]

Jones describes the medical, social, and political understandings of African-American health at that time. Early medical belief held that African Americans were "a notoriously syphilis-soaked race" that either refused to take the disease or its treatment seriously or lacked the intelligence to follow treatment even if they did understand the disease. Although scientists were linking syphilis to African Americans with less inevitability by the late 1920s, many of them still held that syphilis affected the races differently, often believing that African Americans were racially more prone to it.[3] Moreover, by the early 1930s, Public Health Service surveys showed that in some cities syphilis was occurring disproportionately among African Americans, adding further "proof" (although these people were also among the cities' poorest) that it was an African-American disease.[4]

By the end of the nineteenth century, moreover, health officials had begun to grow concerned about the danger of syphilis spreading from African-American to Anglo-American men and women. At the same time, the rise of public health's knowledge and influence drew closer attention to the importance of environmental factors in the spread of the disease, most notably the ill effects of low incomes, particularly in relation to education, sanitary facilities, and access to good or regular health care. Public health officials also often saw themselves as medical guardians of communities, casting officers in a paternalistic role that tended to render them blind to their own biases. Thus, Dr. William F. Brunner, a Georgia public health official who was active in antisyphilis work during the early 1900s, observed, "The negro is here for all time. He depends upon the white man for everything that makes up a civilization. These two statements being true, he is what the white man makes him." Brunner concluded that public efforts toward improving the health of African-American men and women had a direct benefit for the white majority, as, "in doing this we protect ourselves" (an *our* that implicitly places African Americans outside the public community).[5] Thus, although public health's approach to syphilis acknowledged the importance of treating syphilis in African Americans, white self-interest and sense of superiority were often an intrinsic part of such reasoning.

The Tuskegee Study was preceded by a survey of several rural southern communities. In an effort to gain a reliable estimate of the incidence of syphilis among African Americans in these areas, teams of Anglo- and African-American health officials offered demonstrations of the testing for and treatment of syphilis, at which time members of the community were urged to be tested. Drugs for testing and some treatment would be provided free to those communities. Parran hoped that these demonstrations would serve as the beginning of a nationwide syphilis control program.[6]

O. C. Wenger was put in charge of the Mississippi demonstration, the first one of the series. He typified the complex attitudes of progressive public health officers of the time:

> In unguarded moments he was capable of making racial slurs, but he deplored his profession's widespread neglect of the health needs of blacks. He had no sympathy for the misplaced sense of professionalism that prompted many private physicians to oppose public health programs for the poor, and he developed considerable skill at circumventing their opposition. Paternalistic in his dealings with the poor and uneducated, he seemed to enjoy excellent rapport with the lower-class patients with whom he worked in the South. . . . When Dr. E. L. Keyes of New York City, a former president of the American Social Hygiene Association, observed Wenger taking blood samples from blacks on a Saturday afternoon in a crossroad country store, he likened it to "holding high Wassermann in the marketplace."[7]

A somewhat contrary picture, however, is offered by Susan Reverby, a historian who cites a taped interview of Eunice Rivers, the African-American nurse whose role in keeping men in the Tuskegee Study has raised considerable controversy.[8] Rivers recalled the efforts she often made to soothe the tensions created by Dr. Wenger's temper and sharp tongue.[9]

In the days of the demonstrations and treatment programs, drugs were withheld in some circumstances. Although Wenger would have preferred to treat all of his patients who had syphilis for as long as possible, he was forced to concede to the economic pressures of the day, which stated that absolute cure was too costly. Thus, Wenger discouraged the treatment of many older patients whose syphilis was noninfectious and provided only enough drugs to other patients to render the disease noninfectious.[10]

The eventual design of these early syphilis demonstrations clearly served as the prototype for the Chicago Syphilis Control Program, with their publicity campaigns, mass testing, provision of free drugs, and goal of non-infectiousness rather than cure. Wenger was so successful in generating enthusiasm for the demonstrations that he inspired both Vonderlehr and the surgeon general to continue the public show in the national campaign. As Vonderlehr wrote to Wenger, "You and Paul de Kruif have so aroused public interest in the venereal disease problem in this country that it would be impossible to limit the organization to mere health officers. Almost everyone in this country would desire to get aboard the bandwagon."[11]

Before becoming special consultant to Chicago, Wenger worked with Vonderlehr on the first offshoot of the syphilis demonstrations, the Tuskegee Study. Performed in conjunction with Tuskegee Institute, the Tuskegee Study was always a federally funded and run program, with Vonderlehr the first director of the study's fieldwork in Macon, Alabama. He and Wenger headed the team that selected the 399 infected African-American men whose syphilis would remain untreated throughout their lives.[12] In retrospect, it is easy to see the questionable ethics—and miscegenation—of the Tuskegee Study. Such bias makes it difficult to acknowledge the complexities of the evolving social, cultural, and professional consciousness of African Americans at the time; the unique position and relatively revolutionary philosophy of the still-new field of public health; or the relative infancy and naiveté of the field of medical ethics. It is important, although sobering, to realize that people like Wenger and Vonderlehr represented some of the most liberal thinking about race and medicine of their time (or at least among those thinkers with significant political influence).

The Chicago Syphilis Control Program differed from the syphilis demonstrations of the early 1930s in two ways: It involved members of all of Chicago's racially and ethnically diverse population, and it moved syphilis control from a rural to an urban setting. Race continued, however, to be a central issue. In June 1936, Vonderlehr wrote to Frank Jirka, then state director of Public Health for Illinois, "I am sure that you realize that authorities on the control of syphilis appreciate the fact that this disease is a special health problem in areas with a predominantly urban population or where the percentage of Negroes in the general population is exceptionally high."[13] The opening words of the first annual report of the Chicago Syphilis Control Program, submitted to the Public Health Service by Wenger, read, "The city of Chicago has a population of 3½ million people, approximately 10 percent of whom are colored."[14] Despite—or perhaps because of—this acute awareness of race, however, discussions of syphilis and race were few, awkward, and generally unsatisfactory.

At least as early as May 1937 (before the Central Tabulating Unit was in full operation), Ben Reitman was reporting on numbers and conditions of syphilis and gonorrhea in the African-American neighborhoods of Chicago.[15] From the beginning of its tabulations, the Chicago Syphilis Control Program reported its findings by race, with the statistics always showing a higher incidence of syphilis in African-American than Anglo-American men, Anglo-American women always running lowest in incidence for all venereal diseases, and African-American women frequently highest of all groups for gonorrhea. How were those numbers to be interpreted—both fairly and usefully? The skewing of numbers that resulted from targeting populations or locations for testing has already been noted. Officials of the Public Health Service also acknowledged the conscientious response of many African Americans to the call for free testing, which was always proportionally higher from them than it was from other racial groups.[16] Reitman argued for an unbiased analysis of syphilis among African Americans in his earliest reports: "Colored people do not have a preponderance of syphilis and gonorrhea because they are negroes, but because they are poor and live on the lowest economic level. They have fewer baths, less toilet facilities; more of them live in crowded, unsanitary houses. There's more unemployment amongst them and more of them are on Relief and they are less familiar with venereal prophylaxis."[17]

Nevertheless, economic circumstance, created or exacerbated by racial prejudice, combined with that prejudice to perpetuate the image of African Americans as "a syphilis-soaked race" in many people's minds. Ben Reitman fought this notion unflaggingly, always insisting on both the misfortune of circumstance and the pervasiveness of racial prejudice. A few months later, he wrote, "But it is significant that of the 33 positive darkfields 23 were in the colored race and one Mexican, and only 9 white. Something is radically wrong, and humiliating if nothing is done to wipe out the blot that a certain group of our citizens are allowed to have 50 times as much primary chancres as their neighboring groups." Reitman took these numbers as indicative of an economic system that privileged the upper socioeconomic classes, predominantly white. He found support for his analysis in the frustration and anger of Chicago's African-American community leaders, who told him, "Why don't you do something to help us? We represent the lowest economic group. . . . Give us jobs and decent homes, and we'll lo[wer] the V.D. incidence."[18]

Reitman noted the probability that more "colored people are thinking about their health . . . than the whites," thus further contributing to the disproportionate rates between the two groups.[19] He also, and unfalteringly, insisted that poverty and not race lay at the heart of the spread of venereal disease, rejecting the viewpoint of those experts who saw syphilis as an essentially different disease in different races. To this point, he quoted a tribute paid to him by one leader of the African-American community: "We are grateful to you for bringing out the point that our race doesn't have more syphilis, gonorrhea, and tuberculosis because they are colored, but because they are poor and without bath tubs."[20]

Reitman's continued emphasis on the importance of personal hygiene linked syphilis and gonorrhea to poverty and its uncontrollable consequences. He wrote, "And if syphilis is more common in one neighborhood or race than it is in another, it is because they are more ignorant—not more 'sinful' or because of the texture of their skin. Wherever any race has learned the value of personal hygiene and prophylaxis, wherever they have good homes with bathtubs, wherever they understand venereal disease, wherever they have money to purchase soap and prophylactics, there they have a low syphilis incidence."[21]

Reitman expressed anger with other displays of racial bias. In one instance, for example, he complained about a policeman who told a judge, when asked why he arrested so many more black prostitutes than white ones, "It is easier to see the black ones." On the other hand, Reitman slipped into stereotype upon occasion, saying in the same report, for example, "Probably the colored race can handle liquor better than the white."[22] But Reitman's analyses of race and syphilis focused most often on the policies and practices of local and national government that doomed poor African Americans to perpetuate their poor health. "Something is radically wrong," he wrote vehemently after reviewing the monthly venereal disease statistics for September 1937, "if 240,000 colored people have a hundred more cases of syphilis than 3,000,000 whites. Is this an argument for a clinic on the south side or not? If a white man can learn to protect himself from syphilis, why can't a colored man? We should say [because the disparity between women of the two races was even greater], if a white woman can protect herself from syphilis, why can't a colored woman."[23]

The reason, Reitman often clearly explained, was because the information and means for achieving and maintaining health were deliberately being denied them by institutions that would not take the extra steps to redress the injustices their "statistics" uncovered: "It's just disgraceful and humiliating to the ['colored'] race to constantly remind them they have so much syphilis and then not do anything for it," he chided.[24] As Reitman saw it, politics and medicine, science and cultural bias were clearly related. The power differential was clear to him, because the Chicago Syphilis Control Program could pointedly "remind" African Americans of their high rate of syphilis yet withhold what, to Reitman, seemed the easiest and most logical solution.

Before considering the reports concerning syphilis and race that the program issued, the Public Health Service's decision, based on the findings of the "syphilis demonstrations" in the early thirties, that is, to supply physicians and clinics initially with only enough medicine to render syphilis noninfectious, should be noted. Although the program would continue to provide drugs to patients who remained in treatment, physicians needed to request the continued shipment of those drugs and keep careful records of their administration. This policy was based on cost-effectiveness and the goals of public health—respectively, the difficulty of keeping people in treatment for more than a year versus a few months and the historical concern of public health to check the spread of infectious or contagious diseases. The problems of follow-up—keeping people coming back for tedious and uncomfortable treatment—involved not only the expense of the drugs but also (and more costly) the personnel to keep records and, when necessary, retrieve disaffected patients.

This policy, however, had precedent. The same route had been taken in Baltimore in the early 1930s before the advent of the national syphilis control effort. Faced with one of the highest syphilis rates in the country at the end of World War I, Baltimore had begun a vigorous program through its public health clinics. Elizabeth Fee has described the program, which soon became overcrowded and financially overburdened. "The department decided to concentrate on patients at the infectious stage of syphilis," citing as their justification for this policy "that health departments should primarily be concerned with rates of infection, and not with individual cures."[25]

The consequences of Baltimore's policy in 1933 were the same as those in Chicago four years later. Those least likely to be able to continue treatment of their syphilis through its cure, however, were the poorer patients, whose work schedules, means of transportation, and household help were also the most tenuous. Although the concept of mandatory cure also involves issues of privacy, the existing view of African Americans threatening Anglo-American health suggests that the former group's privacy weighed less heavily on the government's mind than the latter group's safety. The Chicago Syphilis Control Program served all ethnic and racial groups, but it was clear from the start that a large part of its work was expected to be directed at poor, urban African Americans. Thus, the consequences of this limited pursuit of cure would inevitably fall most heavily on African-American women and men.

"*The White Race Makes the Statistics*"

The repeated reports of racial differences in the incidence of venereal disease that emerged from the Chicago Syphilis Control Project, often reported in the *Tribune*, soon provoked a direct response to the program's officials in Washington. In December 1938, the *Tribune* printed an article, entitled "Survey Shows Syphilis Drive Barely Begun," which contained the subheading "Most Prevalent among Negroes." It noted that although "Negroes" comprised "only 7 per cent of the city's population, the survey disclosed eight colored syphilitics to every one among the whites."[26]

Surgeon General Parran immediately received letters from a Mr. Barnett of the Associated Negro Press and S. W. Smith, M.D., executive secretary of the National Hospital Association (which represented African-American hospitals). Parran, according to his own account, immediately replied to the two men, then spoke at length with Eugene Bousfield of the Rosenwald Fund, who "came into [his] office" just as he finished the letter, about "the need for better public health service and better medical care among Negroes of Chicago and other parts of the nation without at the same time making more difficult their already difficult part in the modern economic picture."

Parran next addressed these same sentiments to a second person, who had not written to him—John Lawlah, medical director of Provident Hospital, Chicago's African-American hospital. Parran urged Lawlah to convey his support to "the local [African-American] medical society." He concluded, "Up to now there has been the most excellent cooperation from Negro doctors, individually and through their Association, to promote the whole national syphilis campaign. I should be sorry if through misunderstanding of our point of view, or through failure on our part to understand your position, we should cease to be mutually helpful to the cause which interests us both."[27] Enclosed with that letter was a copy of a three-page letter sent to Colonel Robert R. McCormick, publisher and editor-in-chief of the *Tribune*, to firmly correct not the article's numbers but its lack of an adequate explanation of those numbers. Parran pointed out that the statistics were based on clinic attendance rates only and thus merely showed that African Americans had to date responded to the program "much more readily and co-operatively than white or other races"; that syphilis, from an evolutionary standpoint, had spread to African Americans relatively recently and could therefore be expected to be more virulent among that group of people; and that, at the bottom of it all, "Syphilis is no respecter of the size of one's bank account or the pigment of one's skin." Finally, Parran requested another article in the *Tribune* that would clarify its misleading report of December 15.[28]

McCormick complied, and on December 28, the paper announced, "Finds Syphilis Parallels Low Income Groups; U.S. Report Explains High Rate among Negroes." The numbers by race were printed again, but this time supplemented by a statement that "the rate of syphilis closely follows the distribution of population by income classes, those earning the least tending to have the most syphilis." A paragraph heading entitled "Hand in Hand with Poverty" quoted Lawrence Linck: "Among all races, syphilis is the inevitable concomitant of poverty, of ignorance, and of lack of medical care. Where the Negro is at the bottom of the economic leader he is at the top of syphilis prevalence. Wherever his environment is better, his schools better, and good medical care is readily available, the prevalence of syphilis goes down sharply."[29] Lawlah immediately wrote to Parran, thanking him for his letter and for the supplemental newspaper article; he would carry Parran's message to the Cook County Physicians' Association and was writing to Colonel McCormick to express his thanks.[30]

Lawlah's letter expressed gratitude and promised the cooperation that Parran desired. Other African-American leaders, however, were not so easily placated. W. S. Smith's letter to Parran noted that when Doctors

Wenger and Bundesen had asked the support of the Cook County Physicians' Association the summer before, "We were rather leery in entering into the set-up, having realized the very sad experience in a similar campaign several years ago against tuberculosis, the publicity of which proved very detrimental to our group economically." He now felt forced to question their assurance to him that no such damage would occur this time and charged that the Public Health Service "wishes to increase taxation by forcing the few of my group who may be gainfully employed, to lose their jobs and become public wards." Smith was criticizing the privilege possessed by whites, the wealthy class, and government decision-makers. "Why pick on us?" he concluded. "Why not print the ratio with the Jews, Italians, or other racial groups? Why not print the ratio amongst religious bodies? We fail to see any good derived from such publicity."[31]

Parran's response was similar to his letter to McCormick, but he closed with the following advice, which, compared to his words to Lawlah, sound contentious, even mildly threatening: "I am sorry if an unintended injustice has been done; *but if you are as sincerely interested as I am* in the suppression of syphilis and tuberculosis among the Negro people, I think that end can be gained more rapidly by your organization and ourselves working together, each giving the other a fair opportunity to go forward with mutual respect and with an increasing ratio of intelligent awareness as to the whole picture."[32]

The tone of Parran's unsolicited letter to Lawlah is more conciliatory and collegial than his answer to Smith's challenges. The difference might be explained in part by differences in Lawlah's and Smith's actual or perceived influence, or by the history of Parran's personal or professional relationships with the men. Whatever informed Parran's response, however, his words to Smith convey his vision of the U.S. Public Health Service as in charge—and in control—of the design, execution, and outcomes of the Chicago Syphilis Control Program. Paternalism had taken on a sharp edge.

In the aftermath of the two *Tribune* articles, another strong African-American voice in Chicago spoke up angrily. "In America, the white race makes the statistics," the *Defender* observed tersely.[33] The paper charged that fewer whites were known to be under treatment because they went—and could afford to go—to private, white physicians. Chicago's African-American community was clearly watching—and many of them with a highly critical eye—the deeds and words of the Chicago Syphilis Control Program.

The Public Health Service had long depended on—and valued—the support of the African-American community. The syphilis demonstrations in the South had been supported by publicity and money from the Julius Rosenwald Fund, the philanthropic organization created by a builder of the Sears and Roebuck Company. By 1938, the Rosenwald Fund had established an impressive record of leadership in the promotion of health for southern African Americans. In 1940, Wenger credited the fund's support in "initiating" the work of the Chicago Syphilis Control Program, saying, "I repeat and reiterate that the present national venereal disease control program is nothing but an expansion of the Rosenwald demonstrations."[34]

The role of such groups as the Rosenwald Fund or of African-American scientists and health professionals in programs whose racial bias seems clear now is problematic. To dismiss the beliefs and actions of these African Americans, many of whom were middle class, as further evidence of the destructiveness of privilege is, again, to oversimplify the situation. Greater insight might be gained from studying how the liaisons between white political or reform groups and the disenfranchised majority of African Americans were viewed by people like Parran as he wooed, appeased, or rebuked the latter to serve his programs.

It would also be unfair to imply that there was not true, mutual respect between many of the Anglo- and African-American leaders who cooperated on these projects, or that the program's leaders were unconcerned about the fairness with which information about syphilis relative to race was presented. In the face of the challenge of how to discuss important issues, whether in private or in public, however, the Chicago Syphilis Control Program too often opted to avoid discussion altogether, a solution that advances no one's understanding of either race or syphilis. For example, as the displeasure of the African-American community grew, Herman Bundesen's response was to silence any public reporting of venereal diseases with reference to race. "When you tell me how much syphilis there is amongst the Presbyterians, the Baptists and the Jews, we will publish it together with the amount of syphilis amongst the negroes," he stated, an argument that suggests that Smith's words, however impatiently Parran had received them, had circulated beyond the surgeon general's office. Instead of references to race, reports in the *Tribune* began to cite prevalence by city ward. But in a city like Chicago, where neighborhoods were so often racially identifiable, reporting by ward or geographic area would often still enable readers to draw conclusions based on race. By this evasion, the program could deny making misleading, inflammatory racial statements, but little had been done to challenge underlying racial bias.

At first, Reitman acceded to Bundesen's injunction. In early February, he chided the *Tribune* for using statistics based on race, still being compiled by the U.S. Public Health Service, rather than the new, composite numbers now being provided to the press by the Chicago Board of Health.[35] As the disparity in the incidence of venereal disease and the number of cases under treatment continued to grow, however, Reitman found it increasingly difficult *not* to address problems of race.[36] "God Have Pity, the Clouds Become Darker" he titled one of his reports about the continuing grim statistics for African Americans.[37] He began writing about the importance of building trust between black and white communities so that detection and treatment of syphilis could continue. "The average syphilitic doesn't blame himself for his infection: he blames the other fellow, society," Reitman wrote late that summer. "This is especially true in a people who have very little security, joy, or status." After acknowledging the particularly vulnerable position in which most racial minorities found themselves, Reitman went on to describe the tensions that could easily arise and the necessity for respecting people's fears and vulnerabilities: "The south side minority group are unhappy, they are touchy, they have a grievance, they have misery and they are developing a number of suspicions that are not founded. They have the idea that the board of health has been unfair to them and is trying to prove them a syphilitic race."[38] With his references to "the south side minority," Reitman was using racial circumlocutions that enabled those who read his reports to know his racial references, further suggesting that such cautions were not only pointless but also ludicrous. In the end, little was served, at least in the area of public education, by avoiding issues of race in the circumstances and consequences of syphilis.

"Not Allow Any Woman to Hang Around"

The exercise of privilege also influenced the development of policies about women and syphilis. The Chicago Syphilis Control Program actually addressed several groups of women. One group was the "good" women, who were among the first people to volunteer for syphilis tests in substantial numbers. It included women who had played a public role in many of the nation's reform movements or who were privately steadfast in attending to the good health of their families. Bertha Shafer referred to these women when she addressed Mayor Kelley's press conference for the Chicago Syphilis Control Program. "May I tell you," she said, "that there are probably more women in the women's club, more social workers that know about syphilis than there are men, because we have gone on constantly trying—trying to present to them the problems, what they mean to the individual, and particularly the danger of them to the community."[39]

Wife, mother, daughter, or sister as the moral and domesticating influence on the family is a role in which women in the United States have long been cast, by themselves as well as by men.[40] It is a role, then, that was not unusual for the Chicago and national syphilis control programs to call to their support. In November 1937, the *Ladies Home Journal* ran a lengthy story by Paul de Kruif about Chicago's syphilis survey, emphasizing that the women of 1937 "want to know things they never wanted to know before—things beautiful and shocking. Things profoundly true about saving babies' lives and warring on syphilis and infantile paralysis. Things imperative about war and peace and liberalism and what our Constitution meant about liberty." The *Journal* bought a full page in the *Chicago Tribune* to advertise the issue and pictured a woman reading her letter from Herman Bundesen. The headline read, "Yes! I will take a Wassermann Test." The *Journal* praised the women of Chicago—and the history of the *Ladies Home Journal:* "That is what American women are saying today—daughters and granddaughters of women who a generation ago quit the Journal—75,000 of them in a few days—because social disease had been mentioned in its columns by daring Editor Bok."[41]

As admirable an asset as women were seen to be to the program, and as "natural" as their support was deemed to be, as clients they were often viewed with dehumanizing and distrustful paternalism. In January 1938, an editorial in the *Tribune* advocated a state law to require syphilis testing of all pregnant women, similar to a bill that had recently been passed in New York.[42] Such a law was also supported by officials of the Chicago Syphilis Control Program as well as numerous other groups, and when a bill requiring the testing of pregnant women for syphilis came before the Illinois legislature in 1938, it passed with little official comment and even less news coverage—and with no response comparable to the passage of the Saltiel and Graham laws the previous year.[43]

The first dragnet stations were opened in the Board of Health's free clinics, most of which were prenatal facilities used by poor pregnant women, so both the opportunity and the pressure to be tested were unavoidably present from the start. The testing of pregnant women was a cause in close keeping with Herman Bundesen's local and national reputation as a leader in infant health, but Reitman expressed skepticism about Bundesen's motives and priorities. Although he praised Bundesen's accomplishments, he cautioned against a narrowing of the focus of public health in Chicago. "But don't ever forget," Reitman prompted, "saving sick, weakly babies is not the most important thing in the world. If poverty and unemployment is to continue as it has in the past seven years, the babies you are going to save are going into crime and prostitution and the mothers you have instructed will live drab miserable lives and they will rue the day that they ever took your advice and had babies."[44] Although Reitman supported mandatory testing of pregnant women and had made such a recommendation himself in the early months of his work as inspector, he reminded Bundesen of the larger perspective: "It is just as important to save men and women as it is babies."[45] He was quick to detect and confront appeals to public sentimentality about children as rhetorically loaded devices.

Both Reitman's and the Board of Health's statements about the testing of pregnant women, however, reveal mixed feelings about women, their sexuality, and the issue of pregnancy. Reitman's ambivalence toward many of the women he treated for syphilis has already been discussed. His unwavering moral support of women who contracted syphilis was based not only on his belief that venereal disease was not a cause for shame but also on his acknowledgment of women as sexual beings, a view still not widely held in the United States, even after the "sexual revolution" of the previous two decades. Reitman referred to "this 1938, streamlined, cocktail lounge morality," praising the new freedom it gave women—and the strong, brave women who emerged.[46] Moreover, he welcomed them in language echoing that of his past crusade for birth control: "With the disappearance of the fear of syphilis and its twin, the fear of unwanted pregnancy, has come the full development of woman. She has developed a new kind of soul. She is the master of her own body. For the first time in her life she can say truly, 'I will fear no evil. Men cannot hurt me.'" Yet Reitman's argument became more qualified as he continued. "With her development into a cigarette smoker and a cocktail drinker she has come into a new kind of courage. It may be false, but the woman of today stands on her own feet, no longer holding on to the coat tails of father, lover, husband. . . . Because of this the world is going to be different."[47] Reitman's words seem deliberately ambiguous; although he hailed women's loss of the fears of venereal disease and pregnancy, he questioned both the reliability of that courage and the world in which they had won it.

Reitman wrote less ambiguously about another group of women, the employees and steady, unaccompanied customers of the many taverns around

Chicago. Commenting on a newspaper report that taverns were a greater source of venereal diseases than were houses of prostitution, he observed, "That statement is probably true, largely because there are probably five times as many tavern workers and ten times as many tavern patrons as there are prostitutes."[48] He regretted that more energy wasn't being given to arresting the women—he never mentioned the men with them. "ALL THE WOMEN WHO FREQUENT TAVERNS ARE NOT IMMORAL, INDECENT, OR VULGAR," he was careful to insist, but his accusations were still quite broad—"Most of the hostesses, waitresses and entertainers made dates" (i.e., for paid sex)—and his solution sweeping: to test all *women* workers and "not to allow any woman to hang around unless it is known that she has been examined by a doctor."[49] In one report, Reitman gave the names and addresses of fifty-two "South Side Taverns and Other Places Where Women May Be Found."[50] Thus, his attitude toward tavern workers was little different than it was toward women who worked at Emily Marshall's or other organized houses.

Still, Reitman saved his strongest language for prostitutes. "Please, please," he pleaded, "can't we make it plain that whores are women who sell their sex and infected or clean they continue to sell sex and unless they are isolated and kept under lock and key they continue to ply their trade. . . . With tears in our eyes we say again, the whores are not poor miserable unfortunate girls who sell their bodies to make [a] living and support themselves and families and children. Most of the Chicago Prostitutes in 1937 are women with PIMPS, they are vicious, antisocial and disease spreaders and they can not be trusted to cease work while infected, they are non cooperative." More to the purpose of Reitman's bosses, he wrote, "As long as infected prostitutes are permitted to roam about the street at will no progress in venereal control can be expected."[51] A straightforward problem with a straightforward solution—for men.

Also, but quite differently, representative of the period's concerns about the "modern woman" are Reitman's references to "race suicide," the term used for a dominant group's fear that its race would die out from lowered birth rates and increased "defective" births. It was a fear, moreover, that often led to more repressive views on higher education, careers, or delayed childbearing for women, and on homosexuality in men as well as women.[52] Reitman expressed the typical concerns of some members of the white race fearing its loss of numerical superiority when he wrote, "The sorrowful decline in the birth rate and the tragic increase in illegitimacy in Chicago is too serious a question to pass over lightly. The bulk of the illegitimate babies come from the colored race under twenty five years of age. . . . Let's start a birth increase League. In a brief quarter of a century Birth Control has been so successful that it has almost put the human race out of business."[53] Curious words, perhaps, to hear from Emma Goldman's old comrade, but Reitman feared that the rhetoric of syphilis control would deter the general public from having babies.

In addition, Reitman complained that Bundesen's often-stated concern about unhealthy babies did not encourage "potential parents to want children," a deterrent exacerbated by the reluctance of the Board of Health to discuss either birth control or abortion. Some women with syphilis currently under treatment, in fact, did ask what they should do if they became pregnant. "The policy of the Board of Health," noted Reitman, "has been to keep quiet on the subject. Officially it can not advocate birth control or condone abortion." He went on to recommend, "One of the simplest things would be to establish a birth control clinic in the same building as the M.S.H.C."[54] If such a thing could not be done, all women should be advised to go to one of the birth control clinics in Chicago, Reitman felt.[55] He acknowledged, however, that this was "a very touch[y] and dangerous subject" and added, again reflecting his views on the sexual sophistication of women by 1938, that "the syphilitic women and the board of health have grown up intellectually and something will have to be done about it."[56]

A second aspect of the arguments generally used to support mandatory testing of pregnant women was concern for the health of the fetus as opposed to that of the woman. "Syphilis Survey Shows Need for Care of Unborn" heralded the *Tribune* in December 1937.[57] Readers were told that women with syphilis could have healthy babies if they were put under treatment early enough in their pregnancies, a point repeated often in the pages of the *Tribune*. How far was the Board of Health willing to go to protect the health of these fetuses? Reitman recommended registering every fertile woman with syphilis and regularly checking whether she had become pregnant, a degree of surveillance that surpassed current protocol. "Restating a plea made months ago: Every female with syphilis in the childbearing ages should be entered in a special book and the patient should be thoroughly supervised," he proposed, "not by social workers watching their morals and behavi[or] and economic status, but by medical social workers and physicians observing their pregnancies and teaching these women to continue their treatment and to be sure to report to their physicians just as soon as they become pregnant."[58] Although he distinguished between moralistic and medical "supervision," Reitman was nevertheless advocating substantial control over the lives of infected women—a degree of control, moreover, not recommended for men with syphilis who might impregnate uninfected women.

By the time the prenatal testing law went into effect in 1938, 80 percent of Chicago's pregnant women were already being tested voluntarily.[59] Reitman's extensive recommendation was not adopted, but in its annual report for 1938–39, the Chicago Syphilis Control Program announced that "a special drive on congenital syphilis is being inaugurated, in which every effort will be made to secure serological specimens on all women in the child-bearing age, whether they are pregnant or not, and to bring all cases uncovered under care."[60] The plan surpassed Reitman's in the degree of intrusion into women's lives that it proposed and was clearly concerned with babies rather than women, because only women "of the child-bearing age" were targeted for testing. Subsequent reports do not indicate that the plan was ever implemented, but the threat of world war would soon draw the program's attention to women in a different—although not new—way.

"The Heart of the Syphilis District"

Whatever inequities prostitutes, pregnant women, or African Americans faced from the policies and practices of the Chicago Syphilis Control Program, there were other groups perhaps even more discriminated against, but their numbers were even smaller—as were the numbers of their champions. For example, although the lives and health needs of homeless and homosexual men tended to be quite different, they often found themselves at a disadvantage for receiving health care because society did not understand their particular needs; nor, very often, did it seem to care.

Many homeless men were not fully aware of the nature and extent of the antisyphilis drive. Not having permanent residences, they did not receive either a questionnaire or an invitation to be tested, a problem Reitman had pointed out to his bosses early in the campaign.[61] Homeless or transient men, however, were among the first to be mandatorily tested in the resumption of testing by the city courts. Although the incidence of venereal disease among them was fairly high, most of it was "old" and noninfectious.[62] The combination of age, poverty, and homelessness, Reitman pointed out, rendered them relatively harmless to other people. He protested that his talks to them about testing were often "decidedly out of place. . . . when they left they went out to darkness, homelessness.

They were a sexual-less crowd, almost asexual. Hungry men who sleep in shelters some nights . . . 'carry the banner' other nights, sleep in police stations or empty houses other nights are not very much of a venereal menace."[63]

Again, distinctions were being made between infectiousness and cure, distinctions that also implicitly judged a person's worth, but Reitman pointed out that the program's rhetoric was directed to an audience and circumstances essentially irrelevant to homeless men, of whom a special example was made in the press nonetheless. In general, Reitman noted early on that the same message about syphilis would not work for everyone. "If we are going to do effective and efficient educational propaganda," he wrote, "it is essential that we concentrate upon that probably susceptible one-tenth of the population. We know pretty much the types who will contract syphilis, their nationality, and their residence. And we ought to be able to learn how to talk to them or write for them in a language that they will understand."[64]

For other reasons, Reitman felt that homosexual men who sold sexual services also did not pose a great threat. In one guided tour of the "heart of the syphilis district," he stopped at the Cabin Inn, which featured a show by female impersonators. Backstage, the doctor and his guests were told that "none of them had received the questionnaire from Dr. Bundesen but they stated they would be glad to have a blood test. All of the homosexual men actresses are notorious varietarists and continually on the make for trade," Reitman continued but added, "And it is reasonable to suppose that many of them have syphilis, and we are safe in saying that since most of them are constantly exhibiting their bodies and doing public performances, that most of them are under the care of private physicians or clinics."[65]

Reitman's optimism is surprising (and perhaps naive), but nevertheless he was probably more approachable by hoboes and homosexual men and women than were most other physicians of the time. For example, one letter among his papers is from a woman named Marian, who heard him mention in a lecture the names of several lesbian clubs in Chicago. She wrote to ask him for more information because she was "desirous of making friends of [her] own kind."[66] Moreover, Reitman was probably more knowledgable about the venereal health problems of homosexuals than were most health professionals. For example, he informed his superiors that for many hoboes, homosexual men, and male prostitutes the risk of rectal infection was greater than among other groups. "In the hobo district, especially on West Madison Street and also on State street, rectal intercourse is very common," Reitman advised. "All the clinicians who deal with this type of a patient should be careful to include a rectal examination."[67]

He seems to have had little contact with homosexual men who were not in the sex business, however, so his viewpoint was limited. He pointed out that people who earned a living by their bodies were usually frank and matter-of-fact about their sexual practices. Still, he often referred to homosexual activities as "abnormal," at one point espousing a current medical theory that homosexuality was caused by the body's lack of a certain hormone.[68] In one report, Reitman deplored both homosexuality and erotic movies showing homosexual acts, which, he said, would "decrease the birth rate [and] increase the amount of abnormal sex," sounding his fear of racial suicide as well.[69] Similarly, he wrote that it had "been definitely established" that a fear of pregnancy or disease caused a person to forgo heterosexual sex, which in turn could lead "the individual [to] resort to masturbation, homosexuality or a thousand and one forms of perversion."[70]

However open Reitman professed himself to be about homosexuality, his repeated references to "abnormal sex" echo the tone of his occasional outbursts about the "pathological" behavior of prostitutes, further suggesting that his open-mindedness to both prostitution and homosexuality may have reflected his openness to—even delight in—talking about sex in all its expressions. A closer examination of his words may still align him in many ways with the heterosexual morality he so delighted in offending.

Whatever his feelings about hoboes or homosexual men, Reitman never hesitated to search them out and urge their participation in the Chicago Syphilis Control Program. In some ways, he pursued them with even stronger determination, because he saw himself as one of the few health professionals who truly cared about their welfare. He also brought the message of the Chicago Syphilis Control Program to these men in their language, addressing their particular needs. Moreover, with these two groups of men, Reitman's supervisors did not censor his words, as demonstrated in the following soapbox talk he delivered to homosexual men in Bughouse (officially, Washington) Square, a longtime forum for street speakers.

Reitman quickly stole a crowd from Swasey, who was giving one of "his masterful anti-religious talks," by announcing, "We are now going to talk on venereal prophylaxis for the homos." He then proceeded, "in language more frank than is expressed in this report," to exhort his listeners:

> Looking over this crowd we recognize more than 40 men who might be called irregular in their sex expression. In the general population

> from 3 to 5 percent of the people are homosexual. Here in Bughouse Square tonight from 20 to 30 percent of the crowd might be properly designated "queer." This present syphilis drive, the fear of venereal disease, has prompted many people to avoid normal sexual intercourse. With women the fear of pregnancy and venereal disease drives many a fine soul into the arms of another woman. Please do not charge us with inventing any new sexual practices or methods. One can't catch syphilis or gonorrhea very easily from drinking cups, towels or bathtubs, but one can catch it easily by having a rectal contact. At the clinics we see quite a number of rectal chancres. We also find a number of men with chancres on their organ that was contracted during the practice of this method. So, if you men choose this method of expression, be sure to wash your organ quickly and thoroughly with soap and water, and apply a venereal prophylactic tube. It might be well to wear a condom when you are doing this act.[71]

Besides knowing the language to use and the information to give when talking to these groups, Reitman was able to sense their attitudes toward the Board of Health's programs. And what he discovered, paralleling in several ways the attitudes of the African-American community as well, was dissatisfaction with their treatment by government institutions in general and distrust of those organizations' intent or good will. In the case of homosexual and heterosexual prostitutes, Reitman wrote about one publicized raid in a commercial sex district,

> Whenever there are wholesale raids, similar to this one, you can take it for granted that either the Church or Big Business is behind them. In this case, we must give full credit to Sears and Roebuck. State Street between Van Buren and Harrison Street is crowded with men and women, but only a limited number of them go into Sears store. But thousands of male joy-seekers went into the Rialto and to Blondie's and the Dugout and the other places of amusement. Sears put in a rap to the Illinois Vigilance Committee and Bishop Shields put in a rap to the Mayor. The Catholic Youth Organization has its headquarters just a block away. The Catholic League for Decency had its finger in the pie, and there wasn't enough political pull in the first ward to save them.[72]

Reitman reported a similar sense of marginality from hoboes. When he urged the crowd at a gathering in the hobo college to get a blood test, he encountered only anger. "This Syphilis drive is a Racket," one man shot back. Others made similar charges, which Reitman reported: "If you haven't got syphilis they will give it to you." "You have to take 105 shots, 30 in the arm and 70 in the hip and then when you finish your treatment you are worse off than when you began." "It is a graft of the A.M.A. . . . They take the blood test from the poor stiffs, but they never bother the rich or the Society Folks. The Health Dept. is a part of the Capitalistic system who want to exploit and make life hard for the poor and the Hobos." Reitman listened to the complaints, hearing in the crowd's bitter words anger at a program that people felt looked down upon them. At the close of this report, he advised the program's leaders, "More great care and thought must be exercised not to hurt or offend the Hobos and others when they are blood tested."[73]

Although his expressions of racial and heterosexual condescension cannot be ignored, when Reitman wrote about—or talked to—homosexual or homeless men, pregnant women, or African Americans, he nevertheless wrote and spoke as a physician who had been involved in caring for all of these pople and who knew their socioeconomic conditions. Thus, he could often speak for, about, and to them with a special sensitivity to—and sympathy for—their "marginality," another term from a later generation but one that, by his own reactions, Reitman understood clearly. He began with the assumption that men and women were thinking and intelligent and concerned about their health, that they would care even more, or more effectually, were they given the economic opportunities and human respect accorded to most white, middle-class, heterosexual men.

Thus, although it must be admitted that the Chicago Syphilis Control Program did bring new services and opportunities to many people who had previously been denied them, it must also be acknowledged that bias nevertheless played a role in setting limits to the gains made in the overall welfare of many people. Advances in provision of care cannot be abjured totally, but praise of the program's medical accomplishments must be tempered by an awareness of the social limitations that it reinforced or failed to address.

CHAPTER NINE

Paranoia, Prudery, and Profit

Besides invoking legal or moral duty as a reason to be tested for syphilis, officials at times advocated another tactic to recruit participants: fear. According to Surgeon General Thomas Parran, "The only good patient is the frightened patient," and O. C. Wenger celebrated "the prick of the needle, which [students] will forever associate with syphilis."[1] When asked, "Are we not alarming the public to such an extent that they will develop fears which may be as bad or worse than the malady itself?" Dr. Irving S. Cutter, health editor of the *Chicago Tribune*, replied, "It may be a healthful measure, for a while at least, to keep hammering at the dangers of the infection until the machinery can be set up for its control."[2]

Ben Reitman was not willing to accept this reasoning. Instead, he saw it as a cruel tool that could bring shame and despair upon guiltless people. "I suppose," he wrote, "it is true that the reason that some people are moral is because they are afraid of getting a disease. All of this leads me to the conclusion that we must do a great deal of educational work along this line."[3] The term *syphilophobia*, coined in 1914, referred to the deliberate creation of an irrational, unrealistic fear of syphilis.[4]

Reitman considered any kind of fear or ignorance about syphilis to be detrimental to patient, potential patient, and practitioner alike. Syphilophobia could prevent some people from even being tested, their dread of the disease—or its social and economic consequences—causing them to prefer ignorance of their imagined fate. Certainly, testing positive could cause a person to become the object of other people's fears about the disease, however knowledgable or level-headed the infected person might be. Although confidentiality and privacy become important in a society where syphilis is unreasonably feared and people who have it are persecuted unfairly, such secrecy can also have the adverse effect of perpetuating syphilophobia. Similarly, testing groups that are already socially marginal can further exacerbate syphilophobia as well as the existing prejudices against these groups.

Although Reitman saw that these methods often fueled syphilophobia unconsciously, he also believed that syphilophobia was being used consciously, to medicine's benefit. Teaching prevention of syphilis through fear rather than physical protection would not ultimately stop its spread, Reitman argued; it would only feed the coffers of physicians. Thus, syphilophobia was often inextricably both method and consequence of the work of the Chicago Syphilis Control Program. The lines between necessary and unnecessary caution, concern and intimidation, protection and prohibition are often fine ones. Although most of the leaders of the program believed that the dangers of syphilis were urgent enough to justify the use of fear, Reitman persisted in calling foul whenever he felt that the press, the government, or the medical profession had evoked it too freely.

"On the Scrap Heap"

As he surveyed what was being written and said about venereal disease, Reitman concluded that syphilophobia was pervasive and appeared in many guises. One of his favorite targets was the Public Health Institute (PHI), whose "lying sadistic" advertisements appeared regularly in the *Tribune.*[5] One that especially angered Reitman featured a photograph of two babies, clad only in diapers and looking up at someone outside the frame of the picture. The large headline, in capital letters, proclaimed, "THEY NEEDN'T DIE."[6] Reitman charged that such advertisements only fueled the too-easily fed fires of syphilophobia. "For the past few years the newspapers and magazines . . . have presented venereal disease in the most malignant, terrifying, and often exaggerated manner," he complained.[7]

Although the baby photograph may have been the most outrageous of the Public Health Institute's advertisements, it was by no means atypical. For example, on June 14, 1937, shortly after the passage of the Saltiel Law, the institute ran a three-quarter-page advertisement that featured a large picture of a bride and groom. The headline read, "WHAT'S AHEAD? Nothing but misery if either has unsuspected venereal infection."[8] A less melodramatic but still inflammatory advertisement showed a smiling family and announced, "Marriage Test Law in Effect! It's Your Move in the national drive to stamp out the menace of venereal infections!"[9] Although the advertisement urged participation in the new marital laws, it implied guilt on the person who would willfully, by trying to avoid the premarital examination, infect his or her family. Reitman called such tactics a "con" and the institute a collection of "First Class Con Artists."[10] Much preferable, in his opinion, was an advertisement from the United Medical Service (UMS), another low-cost clinic in Chicago, which also ran a picture of a bride and groom.[11] This time, the caption, "Marriage and Health," emphasized well-being rather than the threat of misery and reprehensibleness.[12]

Much of the advertising copy from the Public Health Institute emphasized the innocence of the people who might be infected, and none of the advertisements depicted people of color (nor did other advertising copy in the *Tribune* or most white-run newspapers of the day). Under a photograph of a well-dressed middle-aged woman who is crying, her hair in disarray, the PHI announced, "50 Percent INFECTED INNOCENTLY"; another photograph of a beautiful baby appeared above copy that asked, "BABY COMING? Make sure you have no venereal disease to cause disaster at birth!"[13]

Such images heaped blame on the people who spread syphilis, but other advertising spelled out more plainly the degradation of *all* infected men and women. For example, there was the anguished man clutching his head, with a grimace on his face and the caption, "ON THE SCRAP HEAP"; horses running neck and neck, with the editorial comment, "YOU CAN'T WIN in life's race if you are handicapped by a neglected venereal disease!"; the pointing finger that accused, "BLAME YOURSELF if a neglected or self-treated venereal disease wrecks yoŭr life"; or—one of the few notices to foreground gonorrhea—the man with furrowed brow and sleepless eyes, his head on a pillow, berating himself, "I WAS A FOOL TO TREAT MYSELF!"[14] The most sexually suggestive but deliberately ambiguous advertisement showed a man in a two-piece bathing suit dashing through the waves between two women, also wearing bathing suits. The trio is holding hands. "DON'T LOSE OUT," the headline read, and the text continued, "Gonorrhea or syphilis needn't bar you from sports if you'll procure treatment from your family doctor or this reliable clinic—at once!"[15] Although one might argue that the "sport" indicated is swimming, sexual sport (primarily for the man in the picture) can also be read into the message.

The United Medical Service's advertising followed more the tenor of its advertisement for premarital testing. Thus, on one day readers of the *Tribune* saw a skeleton's hand throwing dice while they were asked by the

PHI whether they were "GAMBLING WITH DEATH?," and just a few pages later the UMS ran an informational advertisement entitled "Eyesight and Health," which included information about the effects of syphilis and gonorrhea on eyesight and also reported the full range of eye examinations and treatment the clinic offered.[16] One advertisement directed specifically to syphilis and gonorrhea pictured a well-dressed man with a dog, whose role as a seeing-eye animal is evident only upon reading the headline: "Syphilis and Gonorrhea Are Curable."[17]

The UMS treated many other health problems in addition to syphilis. As one of their advertisements put it, the service had designed itself to offer a full range of medical services because "no part of the body is safe from attack" in venereal disease.[18] The ability to offer a wider range of services might well have caused the United Medical Service to feel less competition from free clinics and thus to produce less melodramatic copy. During the months of the Chicago Syphilis Control Program, its advertisements announced, "BETTER HEALTH is possible through a greater use of Present Medical Knowledge" and "Cancer Is Curable," with information about syphilis tests and treatment relegated to an inset in one corner of the large space.[19] Reitman generally praised the even-handedness of the United Medical Service's advertisements, but he did record one major gaffe: "AN UNHAPPY CHILDLESS MARRIAGE is almost certain to follow if either should have a venereal infection" ran during the early days of the new marriage laws.[20]

Even before the advent of the Chicago Syphilis Control Program, however, the PHI had run sensational invitations to Chicagoans to be tested or treated for syphilis. Both the institute and the UMS had advertised regularly in the *Tribune*, the PHI about once a month and the UMS not quite as frequently. By the summer of 1937, the PHI ran one-third- to three-quarter-page advertisements almost weekly, still about two to the UMS's one. Typical of the advertising of the United Medical Service before 1937 is a brief article on the importance of dental hygiene, which was accompanied by an outline of its own dental services.[21] For the PHI, advertising showing smiling brides and grooms alternated with more sinister images, such as a man in black hat, black coat, and black mask, with the banner "A MASQUERADER! Is this dreaded disease that appears in many different disguises."[22] The image, which portrays syphilis as evil rather than the person with syphilis as the guilty one, invokes syphilophobia in its truest sense because it focuses fear directly on the disease itself (although still portrayed in human form). Reitman's anger with advertising like that picturing the two babies, however, illustrates the short distance between generating fear or hatred of a disease and fear or hatred of the person who has that disease.

There is no information about whose advertisements generated more tests or brought more people into treatment, but Reitman reported several times that the PHI's had not succeeded in bringing in large numbers of new patients. In fact, their business grew increasingly bad. "It Costs a Hundred Dollars to Get a Case of Syphilis and a Thousand Dollars to Get a Primary Chancre," Reitman observed about the diminishing returns of their advertising campaign.[23] He offered no comparable figures for the UMS but insisted that the PHI was getting what it deserved.

When the PHI ran an advertisement asking "Am I Really Cured?," Reitman charged that such a message fostered fear that would lead to further, unnecessary treatment. He held that such advertising was "not only bad taste and exaggerations, but they [were] deliberate falsehoods" and predicted that they would only "result in frightening the public, driving them into the very things they claim syphilis is responsible for: nervous breakdown, and even insanity."[24] Such irresponsibility, he believed, should be punished. "Any advertising pay clinic who runs such sadistic ads as have been appearing in Chicago for years should be censured," Reitman counseled.[25]

Reitman, however, did not limit his criticisms to the Public Health Institute. "Not only should the P.H.I. be asked to soft-pedal syphilis, but everybody else who is doing syphilis control work" he recommended, referring to exaggerated claims about the consequences of the disease.[26] The press was also guilty of dangerous exaggeration, and Reitman often attacked the sensational headlines, which he called "scareheads."[27] For example, while he praised the officials of the Board of Health and the *Tribune* for one of the "best and one of the most accurate pieces of publicity that have appeared so far," he still chided Bundesen for making the unsubstantiated statement, in that same article, that "a great number of the city's 11,555 heart disease fatalities, which lead all others, were undoubtedly due to syphilis." Reitman cautioned Bundesen to use his authority with more care, because he would "be often quoted. . . . We have so many reliable, accurate statistics," he concluded, "that it is just too bad to spoil a splendid story with a haphazard guess."[28]

Another time, Reitman attacked an article in the *Journal of the American Medical Association* as "the most terrifying, exaggerated, stupid article on syphilis that has been read for a long time," leading him to wonder, "What Is There in the Soul of the Average Syphilographer and Health Officer

That Wants to Make Him Exaggerate?"[29] Intent as Reitman was on bringing all cases of syphilis under treatment, he abjured such exaggerations, believing fear to be a greater deterrent to honest discussion than it is a spur to action. For example, he returned from giving a lecture to seventy-five women students at the Pestalozzi Frobel Teachers College in despair, disbelief, and anger. He wrote, "Although these women were educated, intelligent women their questions appeared dumb, naive. The girls were plainly frightened and horrified. . . . Where do these girls get their ideas about syphilis? If we may be allowed to say so, unhealthy and misinformed ideas? They got them from the Chicago Tribune, from the Ladies Home Journal, from various lecturers at the Women's Club and the YWCA and from hearsay and from ads in the washrooms."[30]

One of the more ironic examples of syphilophobia is a series of stories that ran for a week in February 1938, just as officials of the Chicago Syphilis Control Program were preparing to open the free testing centers. "Test Sanity of Al Capone," the *Tribune* announced on its front page, "Gangster Reported Ill of Paresis." Unconfirmed rumors had it that the famous gangster had collapsed "possibly as a result of a dread mental disease." Chicago detectives said that they had never heard that Capone had "any syphilitic infection," but they recalled "that he started as a proprietor of vice resorts on the south side. More than fifteen years ago, when these places were raided, employees of the health department found some inmates infected with syphilis." His condition was described even more vaguely, but much more picturesquely, by fellow prisoners, as word spread from cell to cell that "Capone's cuttin' up dolls."[31] For the next few days speculation about Capone's diagnosis ran high, but his personal physicians repeatedly denied the possibility that his breakdown was caused by "paresis or from any venereal disease."[32]

Capone's story is ironic in several respects, the most obvious of which is the association of syphilis with crime that kept appearing, deliberately or not, in the pages of the *Tribune*. Second, whatever Capone's diagnosis might have been, that he had run a prostitution business was in many people's minds proof enough that he had syphilis himself, again linking syphilis to public standards of immorality. Most pertinent, though, is the eagerness of Capone's physicians to absolve their patient of the onus of having syphilis, suggesting that the conviction of having syphilis was worse for his public image than other crimes of which he might be guilty.

As with the equation of syphilis with race rather than poverty, using the threat of syphilis to evoke fear and blameworthiness rather than a conscientious nonjudgmental concern would only foster traditional prejudices against syphilis and, by association, against people who were infected. Such strategies, although perhaps prodding anxious people to go to their physicians or a testing station, risked failing to teach them either to understand syphilis or be compassionate toward those who contracted it. Assessing the effects of syphilophobia on a national level, the historian Allan Brandt points out that the psychological effect of syphilophobia was, in fact, the opposite, that "the burden of syphilophobia . . . often generated unrealistic views of the disease, its impact and treatability—attitudes that ultimately made syphilis a more difficult problem to handle."[33]

"*Make Sex Safe*"

One of the reasons that Reitman eschewed the use of scare tactics was because he felt that an irrational fear of syphilis would instill an equally irrational fear of sexuality itself. In addition to his concern that fear of syphilis could lead to a decrease in population Reitman also suggested that syphilophobia could lead to homosexuality or masturbation as a substitute for vaginal heterosexual intercourse, both of which he considered to be perversions of the sex urge. Reitman saw sexual prudery inherent in syphilophobia, a squeamishness that made the forthright discussion of any sexual matter taboo. How could the taboo surrounding syphilis be lifted if sex was given no public forum? Only when people were ready to talk realistically, without embarrassment or censure, about sex, Reitman believed, would they be able to deal decisively with syphilis and gonorrhea. And "realistically" meant acknowledging that many people would continue to have sex, outside marriage, with multiple partners, whatever the risk. "We must make sex safe, fool-proof," Reitman insisted, adding, "And this is a comparatively simple procedure."[34]

Reitman frequently criticized the sex education that the program sponsored, which most often took the form of "short, informative talks" at factories, businesses, clubs, and schools. Its main purpose was proclaimed to be to teach "the value of serological tests," and talks were always followed by the collection of blood specimens.[35] The talks generally emphasized the pervasiveness and devastation of syphilis, with statistics on how the rates of insanity and cardiovascular, neonatal, and neurologic deaths could be vastly reduced by treatment or how much money in income, salaries, or goods produced were lost to syphilis annually. The education and testing of schoolchildren, a much-publicized activity of the Chicago Syphilis Control Program, was cautious. "School authorities," wrote one *Tribune* reporter, "explained that tax-supported education must not get too far ahead of public opinion for fear of an unfavorable reaction. According to educators, parents in general have not shown a readiness to have their children receive sex instruction and are even less inclined toward teaching the nature and dangers of syphilis."[36]

By March, Chicago had received nearly $4,000 from the Public Health Service for public education, and officers of the Board of Health and the Illinois Social Hygiene Association were spreading the word. In one article, the *Tribune* reported that Board of Health Secretary Louis Schmidt would be addressing a hundred members of the clergy the next day, and Dr. Bertha Shafer, an obstetrician from Northwestern's medical school, would speak to eight hundred members of the National Youth Administration at a special meeting to be held at Hull-House that coming Thursday.[37] Reitman also addressed many young people and adults about venereal disease, but he never shared the podium with Shafer or Schmidt. Although he spoke to some of the same groups they addressed, the roster of his speaking engagements is much more varied than theirs. In the period from January 1 through mid-November 1937, for example, he reported speaking to more than a hundred audiences—American Legion posts, YMCAs and YWCAs, the University of Chicago, Northwestern University, Baptist and Methodist churches, the Sociology Club, Criminological Society, Chicago Federation of Labor, anarchists, socialists, communists, the Boys' Brotherhood Republic, hobo colleges, unemployed councils, Workers Alliance, Bug Club, Workers Colored Forums, the Plebeian Forum, Atheist Society, Seven Arts Club, "the sexeteers," Artists Equity, the Purple and Gold Club of the University of Chicago, and Harrison, Crane, and Englewood high schools.[38]

Reitman's talks were a natural extension of the presentations he had given for years to the hobo college and the anarchists, from his soapbox at Bughouse Square, or to the often raucous literati assembled at the Dill Pickle Club.[39] His "crime tours" also long predated the Chicago Syphilis Control Program, as he volunteered to colleagues or hired out to various civic and youth groups to introduce them to the people and places of Chicago's transient, bohemian, and sex districts. The format, though, was easily adaptable to the arena of syphilis control, and Reitman was happy to accommodate it. The following description is typical of one of these tours:

> The number of students who are studying sociology, criminology, and vice are constantly on the increase. It was a joy the other night to have a group of mature, sincere, University students to take on a tour of the West, North and South side vice areas. If the town is tight or closed up, there's no evidence of it. Prostitutes can be seen in large numbers in all parts of the town. Street walkers, both black and white, are more common in Chicago than for a long time. Window tapping can be heard on both the West and South sides. Street walkers are common and brazen. The whores are verbose. Invite them to have a drink and they'll talk about their work. Homosexuals ply their trade on the North and West and South sides in the open. Sex—sex—everywhere.
>
> The University students asked a lot of questions. "What percent of the prostitutes are infectious?" "The Board of Health statistics show that about fifty percent of them have either gonorrhea or syphilis or both." . . . "Why doesn't every man who stays with an infected prostitute get syphilis or gonorrhea?" "Not one in a hundred men who contact prostitutes catches a disease from her." . . . "Why?" "The man washes his organ and voids and takes a prophylactic. Often he will wear a condom, and although many prostitutes have venereal disease, they are not very infectious."[40]

Reitman's syphilis talks differed markedly from those of other officials of the Chicago Syphilis Control Program in two significant ways. First, as Reitman had agreed, they were not presented as being under the program's auspices. Second, Reitman's talks always included information about the prevention of venereal disease by other than "moral" measures.

From the beginning of his work with the Chicago Syphilis Control Program, Reitman had complained about the inefficacy and irrationality of moral prophylaxis. In May 1937, he attended a luncheon of the Committee of Fifteen (Chicago's long-standing antivice committee), which was visited by Health Commissioner Rice of New York City. Reitman reported that he

> butted in and asked, "Please, Mr. Commissioner, will you tell us why Health Commissioners don't push venereal prophylaxis?" His answer was, "I don't know a good prophylactic." We suggested soap and water and voiding might do some good. We only had a minute and a half to talk going down the elevator but with all the power God gave to us, we pleaded for venereal prophylaxis, ending with "Unless you health officers teach and urge prophylaxis, you will always be the servants of

> the sacrament of the needle and reformers." We were unable to tell whether the look in the eye of the President of the Chicago Board of Health as he rescued his colleague from New York and hurried him down the street was one of pity, contempt or anger.[41]

Reitman received similar responses in subsequent months as he tirelessly repeated his petition. He regularly received two answers. One was medical: "There are no reliable prophyla[ctics]," Assistant Surgeon General R. A. Vonderlehr said in a private conversation. Reitman agreed that some condoms were faulty but that their manufacture could be standardized. In response, however, Vonderlehr countered with the second argument: "The members of the Congress and the Senate and the Woman's Clubs and the Educators would be down on our necks if we attempted to advocate prophylaxis."[42]

The methods of mechanical and chemical prophylaxis for venereal disease that existed in 1937 and 1938 were essentially no different than those that exist today. Condoms, the method most widely known at the time, were less reliable than they are now because of the lack of enforceable standards that Vonderlehr had referred to in his conversation with Reitman. In the late spring of 1938, the Chicago City Council attempted to pass a bill that would create standards for "rubber goods," meaning condoms specifically, but the mayor refused to sign the bill. Board of Health officials were trying, through numerous arrests of pharmacists and drug store clerks who sold defective condoms, to force more responsible production, but pharmacists and "rubber goods manufacturers," Reitman reported, "have pooled their interests and are making a heroic defense" in these cases.[43]

Also available for more than twenty years were "prophylactic kits," developed during World War I and usually consisting of a chemical cleanser for washing and flushing the penis plus a medicated ointment to apply to it after washing.[44] A version of the kit had been proposed for inclusion in the fee a man paid for a prostitute during the Chicago Board of Health's earlier antisyphilis campaign in the 1920s. Reitman's goal, as director of the Chicago Society for the Prevention of Venereal Disease, was to develop a family prophylactic kit, and he was "hoping to get a universal distribution of ten cent prophylactic packages in all 5 and 10 cent stores, grocery stores, vending machines and so forth."[45]

The Public Health Service was not, however, totally ignoring the possibilities; the agency had undertaken at least one study in search of a chemical preventive.[46] It had also been reviewing the various prophylactic kits that had become available on American and foreign markets since World War I. The Health Service was not impressed, however, and when Wenger requested that a special group be appointed to study prophylaxis, Vonderlehr responded that he "could see no possible advantage . . . at the present time [because all] of the experimental and practical work which has been done in this field indicate that the methods are not of particular value."[47] It would not be until 1940 that the Health Service would support the use of condoms and prophylactic kits, a position that other organizations adopted as well.[47] Dr. Herman M. Soloway, chief of the Division of Venereal Disease, Illinois Department of Public Health, commented, "We wish to advise that it is the policy of the Illinois Department of Public Health not to publish any articles on prophylaxis."[49] Walter Clarke, executive director of the American Social Hygiene Association, wrote to Reitman that although his organization did not oppose chemical or mechanical prophylaxis, "The simple fact seems to be that public leaders are not ready for prophylaxis as a part of a public health campaign."[50] Finally, Reitman castigated Richard Finnigan, editor of the *Chicago Times*, for being "a strange mixture," courageously printing stories about birth control but refusing to mention prophylaxis when syphilis was discussed.[51]

Although Reitman talked about condoms, creams, and soaps for prevention of syphilis and gonorrhea, soap and water remained his most frequently touted prophylactic. It had been a part of the standard advice to soldiers during the Spanish-American War and World War I.[52] It was a simple procedure: Men should urinate after intercourse and wash the penis with soap and water; women were advised to douche with soapy water and wash their genitalia as well. Of all methods of prophylaxis for venereal disease, this was undoubtedly the most "unscientific," the most suspect in a world that had already entered the age of chemotherapy. This method, however, resonated comfortably with Reitman's own, more anarchistic approach to health care, where low-cost, accessible means of prevention counted greater than high-priced restricted means. Easy immediate measures that had moderate-to-good efficacy were, to Reitman, better than no measures at all. A counter-argument, of course, is the danger of instilling false security in the minds of the people using the soap and water, but Reitman would have responded that most people were going to behave sexually whatever the risks. If they could not—and Reitman would have said that they should not—be frightened into abstaining from sex, then they should at least have enough information to begin to make responsible decisions.

"The Public's Reaction"

While many people appear to have welcomed Reitman's prescription, others objected loudly to his proselytizing. It is not always clear whether they were objecting to the message or to the messenger. "The public's reaction to 'soap and water' differs greatly, and the opposition comes from the most unexpected sources," he observed in a report on his talk to the Advertising Men's Post, No. 38, of the American Legion. Adult men, in the harsh, competitive world of advertising, Reitman assumed, "would take a discussion on sex very good-naturedly." Instead, the post damned his presentation. "His viewpoint is pretty much a key-hole philosophy, seen through a West Madison Street flophouse," wrote one legionnaire in the group's newsletter. "He took the cover off of the garbage pail, and during the exploitation, used many naughty words. He caused a few blushes, and called a spade a bloody shovel. All in all, his talk was amusing and Utopian via birth control, sterility and other articles of Sunday Magazine supplement news."[53]

Reitman reported a similar reaction to his talk at the Cook County Medical Society banquet, where "many of the doctors expressed amazement and even horror that one could talk so freely about washing the organs with soap and water." In contrast, he boasted of the openness and appreciation of boys from the Lincoln Belmont YMCA, who "praised and thanked" him "for saying the same things that shocked the American Legion and the colored doctors." Similarly, he described the policemen and detectives at the Des Plaines Street Police Station, who "all laughed and kidded and thought the talk was very amusing. Most of the audience who have heard us say, 'wash it with soap and water' have taken the message seriously."[54]

Reitman's language and style often elicited criticism from his medical peers and bosses as well. "Dr. Louis Schmidt recently warned us not to be so vulgar and sexy," he reported with an almost smug exasperation early in 1938, "and [not] to talk so plainly. He said, 'Sometimes you write interesting things and we want to show them to people, but much of it is so rotten and contains such terrible words that we would be ashamed to show it to anybody else.'"[55]

In May 1938, Vonderlehr wrote to Wenger: "Indeed, it is my impression that the activities of some individuals and their melodramatic and somewhat obscene descriptions of experiments which have no scientific basis may be the cause of the loss of public interest in the present campaign. Certainly it has done no good."[56] It would be hard to prove that Vonderlehr was writing about Reitman specifically, but a few months earlier Reitman had reported on an informal experiment in which he had distributed four jars of "a special mercury ointment" to four women working in different houses of prostitution. After sixty days, three of them reported that most of their customers had accepted the treatment with "no complaints"; the fourth had lost her job for unrelated reasons and left town. Reitman also distributed two dozen tubes of an ointment to men whom he knew to have multiple sex partners, and instructed them in its use, with no complaints other than that it was "a little bit greasy and the tube leak[ed] when carried in the vest pocket sometimes." In fact, he found them generally "glad" to have the ointment and asked to run a larger survey with ten thousand tubes of the cream but was not given permission to conduct the formal study.[57] Whether or not news of the project reached Vonderlehr's ears, Reitman's actions would certainly have fallen into the category the assistant surgeon general had criticized.

Reitman knew that he was pushing the conditions of his continued employment with his statements on prophylaxis, even when he spoke as director of the Chicago Society for the Prevention of Venereal Disease. Although at the start of the program he was willing to distance his work in prophylaxis from his official duties, it often cost him dearly. As he reported on the annual meeting of the Illinois Social Hygiene League, "No mention was made of the fact there was a man in the audience who had all the vice facts at the tip of his tied tongue. This encyclopedia of vice, crime and rackets had been instructed to speak only when he was told to speak." Reitman not only wanted to be heard, but he also had useful information that other people didn't have. In the same report, he conceded that he hadn't always played by the rules set out for him: "We admit that we haven't always been enthusiastic for the Chicago Plan, and we have overstepped our bounds in trying to push Venereal Prophylaxis. We have tried to be loyal to our bosses and our job, and sometimes our missionary zeal has made that difficult to believe."[58]

Still, Reitman desired to play a more visible role in the program, a place to which he felt his accomplishments entitled him. In the wake of the popular reception of his book *Sister of the Road*, he wrote to his bosses, "And now that my sphere is being enlarged and we become an author of some importance, with a vague possibility of hitting the movies (or the big time circuit) we express a desire, a wish to continue our work and to be given tasks that we might do with credit to the department."[59] By 1938, however, Reitman's attempts at polite requests became harder for him to sustain.

"As the months go on, we talk more freely—perhaps more vulgarly—about prophylaxis," he wrote in September.[60] One of his most public outbursts occurred when he appeared, uninvited, at a banquet at Chicago's Provident Hospital, where he confronted the guest of honor, Dr. Reuben Kahn. "Kahn," Reitman accused, "you're a great help to these colored doctors. You're finding so much syphilis that they can buy new 1938 cars, permit their wives to wear seal skin coats. The doctors are buttering their bread with secondary syphilis, taking trips south with the proceeds of neurosyphilis." Dr. Kahn reportedly responded "good-naturedly" but answered with the need to first diagnose and treat all current cases of syphilis. "And besides," he challenged Reitman, "what kind of prophylaxis would you use for a woman?" Reitman seized the opportunity, and "with a plate full of chicken and sweet potatoes we squatted down and demonstrated the two finger soap and water pan method."[61]

Despite this astounding performance and his attack on the economic motives of African-American physicians, a few days later Reitman was invited by Dr. Bousfield to a special meeting of health leaders, both black and white. According to Reitman, "Dr Bousfield watched me carefully, and said, 'If you don't make any demonstrations, and be reserved in your speech, you can tell us a little about prophylaxis.' " He discovered that one of the guests, Dr. MacLean, dean of the University of Chicago Medical School, had been considering teaching prophylaxis to his students. "But what do you think they'd say about the University of Chicago," he asked Reitman, "if they knew we were teaching prophylaxis to our students? They would charge us with teaching immorality and say we were encouraging promiscuous intercourse!"[62]

MacLean's statement is probably more or less correct. Most medical schools at that time taught little or nothing about preventing venereal disease. In his report, however, Reitman responded with only brief enthusiasm for this apparent advance in his cause, launching instead into a tirade against the social forces that contributed to such antedated and prudish medical curricula. He recounted his ensuing lecture:

> A long-haired man undid a dirty finger at the speaker and almost screamed, "Do you hear what MacLean said? That at the University of Chicago they'd rather have their fine boys and girls get syphilis than offend Mrs. Grundy.[63] Not science, or philosophy, or the truth or religion dominates the University—but MOTHER GRUNDY!" DEAR OLD MOTHER GRUNDY, WHAT A POWERFUL FACTOR YOU ARE IN THE

> WORLD. YOU'VE GOT SURGEON GENERAL PARRAN "BUFFALOED." Louis Schmidt and Herman Bundesen who defied the Chicago Medical Society and the World and came out for modern scientific clinical service for the low-salaried group are afraid to offend you! The colored doctors, who see syphilis about to decimate their race bow their heads in reverence before you! But, Mother Grundy, old girl, you can control the health officers and the philanthropists but you can't control the balls and the ovaries of our young people! Dr Embree and Dr Bousfield are converted. Venereal Prophylaxis will become an important part of the Rosenwald Health Program.[64]

Given that Reitman's account could have been embroidered considerably in the retelling, two points can still be made. First, his political—and polemical—strategies may often have put off even people who were sympathetic to his ideas. Second, and more important, this reported meeting, whatever its outcome, documents that prevention of venereal disease was being taken seriously in some fairly influential quarters, whatever the official position on the subject might have been.

About two months after the meeting with Bousfield and MacLean, Reitman reported another meeting with MacLean, who allegedly told him, "If what you say [that the University of Chicago did not teach venereal prophylaxis to its medical students] is true, I will see Dr. Becker, Professor of Syphilis, this morning and find out why. I was in the Army during the war and I know what prophylaxis can do. It can prevent venereal disease and our students can be taught it." Reitman again reported that he spoke to MacLean as the director of the Chicago Society for the Prevention of Venereal Disease, but stressed his sense that there was a growing support for prophylaxis. "Please," he concluded, "for the love of Metchnikoff [the Russian bacteriologist who, working in France, identified the spirochete], understand this simple thing: That if the Board of Health will here and now, immediately, advocate, sanction, at least condone it, they will save themselves a great deal of criticism and possibly trouble."[65]

Reitman did not report any subsequent changes in the curriculum of the medical school of the University of Chicago, but once more he chided the false, misplaced modesty of men who, he knew, had supported prophylaxis in the past. That he was permitted by Bundesen and Schmidt to speak about such matters, albeit without the imprimatur of the program, must have only fueled Reitman's sense of their hypocrisy. Arguments about the lack of a fool-proof prophylactic rang hollow to a man who felt that sexually active men and women were capable of understanding both sex and syphilis well enough to weigh the risks of the various preventive means available and make their own informed decisions.

"More Patients, More Hospitals, More Syphilis"

Not to be overlooked in the spectacle of Reitman's demonstration of the soap-and-water method of prophylaxis for women at the banquet at Provident Hospital was his vivid verbal picture of physicians "buttering their bread with secondary syphilis."[66] He registered his political analysis of the Chicago Syphilis Control Program early on, writing to Vonderlehr, "As I went into Sociology and Economics I slowly realized that the competitive system was highly advantageous to a large portion of the American People and that is why they hang on to it, but for the life of me I can't understand why so many honest scientific socially minded medical men oppose prophylaxis or at least are indifferent or silent regarding it."[67]

The economics of syphilis, by Reitman's reasoning, was related in part to syphilophobia. At one point, he accused the PHI of running fear-inducing advertisements because they needed to bolster their own flagging business.[68] More often, he wrote of the connection he saw between medical economics and moral prophylaxis, with economic gain serving as both an impetus for and a consequence of this position. In a program that emphasized treatment over prevention and whose treatment, moreover, required protracted care and countless visits to the doctor, the potential income from the expected surge of patients might be considerable. "These men unconsciously are like businessmen," Reitman described the officers of the Chicago Syphilis Control Program to the University of Chicago's sociologist Herbert Blumer.[69]

Reitman, consistently wary of capitalist tendencies to economic greed, however, found himself caught at times on the horns of an ideological dilemma. On the one hand, he gloated at the success of the city's clinics, which made no profit from others' illness. For example, he commented with a sense of satisfied irony on the empty waiting room at the United Medical Service, "the most magnificent pay clinic" in Chicago, which sported the "most modern, large comfortable chairs, soft thick carpet," while the outdated Municipal Social Hygiene Clinic had only "rough benches there, but they were all filled, all the treatment booths were crowded and there were several hundred persons waiting to get in."[70] Reitman also noted the loss of patients and income to private physicians, who, health officials had thought at first, would benefit from the passage of the Saltiel Law. "It is just an idle dream for the majority of doctors to think they can make a living out of these diseases," Reitman derided them.[71] On the other hand, he felt that free clinics held an unfair advantage over self-employed physicians. This position, generally not in keeping with his usual support of the program's broad-based methods, may reflect an empathy for private physicians that Reitman, himself a struggling "clap doctor," might easily have felt. Whatever the impulse, from time to time he reported concern for the flagging business of private doctors and low-cost clinics.[72]

At times, Reitman criticized the Chicago Syphilis Control Program for being insensitive to private doctors. He reported on a meeting of Chicago's physicians who specialized in venereal disease, with this comment, "The attitude of the Board of Health in this present syphilis drive was held responsible for the poverty and the misery of the V.D. doctors. . . . Great pains must be taken to prove to the Chicago physicians that the bulk of our patients are unemployed, on relief . . . and there are but very few patients—hardly 5 percent—who could pay the physicians any kind of fee."[73] On another occasion, he commented, "Many of the private physicians and pay clinics think the Board of Health is robbing them of their patients, and the larger the clinics become, the more antagonism there is apt to develop."[74] Some of the tension was reduced with the Health Service's decision to give neoarsphenamine and bismuth upon request to all private physicians (who could still charge their patients for the office visit), who consequently began sending in more smears and blood samples to be tested.[75] But the increase soon leveled off, still far short of the desired level of participation, and the program eventually took on a more active role in testing and treating precisely because it had not received the hoped-for cooperation and support from private medicine.

More often than sympathy, though, Reitman expressed dissatisfaction with private physicians' and low-cost clinics' lack of involvement in the antisyphilis campaign. Although they may have done well in treating venereal disease, he charged, they had been content merely to treat it, not work to control its spread. "The majority of syphilitic patients diagnosed in 1938 did not voluntarily seek the doctor or the treatment center," he argued. "They were located through the efforts of the Federal, State and Chicago Health Departments."[76] He challenged the profit motive of self-employed practitioners because, he believed, "as long as private physicians had control of any disease . . . that disease flourished."[77] When pressed on the subject, however, Reitman extended his accusations beyond private medicine. "No medical institution, doctor, hospital or insurance company has the slightest desire to prevent and eliminate syphilis and gonorrhea," he raged. "They are just as anxious that syphilis flourish in the world so that they may treat it as bankers are to earn interest; real estate owners to collect rent; or business men to collect profits."[78]

The proof of his accusations, in Reitman's mind, lay in the continued refusal of these groups to act publicly upon the knowledge that venereal disease could be prevented by mechanical or chemical means. He tied that position to the unforgivable practice of syphilophobia. In an angry report entitled "You Might as Well Kill a Man as Scare Him to Death," Reitman wrote, "Nowhere in all the literature and lectures is it emphasized that syphilis is unnecessary and preventable. And nowhere is there an ideology directed towards the prevention and elimination of syphilis and tuberculosis and other diseases. In all the colossal plans which are being formulated by government and private agencies the main idea is: more patients, more hospitals, more treatment. And that means more sickness, more syphilis."[79] Reitman's anarchist past echoed through his charges. "Business and industry are run for the benefit of the business man," he said in another report. "It is a profit system for those who control it. It is the same with medicine. The doctors developed medicine and organized it and the sick belonged to the doctors." Emma Goldman couldn't have said it better.

But when Reitman's attention shifted to the relation of government to medicine, his orthodoxy faltered. Rather than deplore institutional authority, Reitman hailed the entry of the government into the regulation of medical practice. A few sentences later, he said, "In Chicago, [private physicians] have been monkeying around with syphilis for 100 years. There was no control of syphilis, there was no diminution of syphilis and things went from bad to worse til 5 percent of our people have syphilis. . . . Therefore it was high time that the government stepped in. The medical organizations have always been the enemy of health departments. In New York, California and other states the 'medical trust' have had the health departments eating out of their hands."[80]

Reitman's opinions about the role and attitude of private physicians in the Chicago Syphilis Control Program were complex and often confusing. To some extent, they were inextricable from his personal beliefs regarding prophylaxis and his idiosyncratic social and economic theories. On the other hand, his words also reflect a basic conflict between the practice and ethic of public health and private medicine. A reluctance to privilege the one over the many, especially when the one is at an economic advantage over the many, characterized not only much of anarchist philosophy and Reitman's medical career but also the impetus behind the creation of the Public Health Institute itself, which, paradoxically, Reitman now often saw as one of those privileged few. Message, motive, and method were inextricable to him. Any message less straightforward than "syphilis is preventable" led him to question the messenger's motives and methods.

CHAPTER TEN

Last Straws

It would have been difficult for readers of the *Chicago Tribune*, during the summer of 1938, to know exactly how the work of the Chicago Syphilis Control Program was faring. Newspapers and city officials continued to announce the defeat of syphilis, but behind the scenes there was growing concern about the continued efficacy of mass testing, mounting tension around the handling of information with regard to race, and more frequent calls for mechanical and chemical prophylaxis. By late summer, it became increasingly clear to people close to the program that enthusiasm for it and its activities had peaked. In addition, program officials were encountering new problems in funding their work as well as shifting public opinion toward some of the federal programs that backed the antisyphilis campaign. Finally, there was a new, increasingly urgent, demand on public attention: the growing threat of world war. The program was changing, both deliberately and consequentially. As it moved into the 1940s, it underwent an incremental and unheralded dissolution.

A Delicate Balance of Powers

The Chicago Syphilis Control Program would never have come into existence without the impetus and support of the federal government, particularly the programs of the New Deal, a relationship of federal, state, and local agencies that threw into uneasy alliance a collection of often contentious offices and individuals. At the end of 1937, Ben Reitman observed, "How fortunate Commissioner Bundesen, Secretary of the Board of Health Schmidt, and the Director of the Social Hygiene Division Taylor were to be able to secure the financial aid and cooperation of the U S Public Health Administration, The Works Progress Administration, and the Illinois State Board of Health."[1] Whether he spoke with ardent appreciation, naive optimism, or prophetic irony, Reitman was correct in viewing the program as a grand experiment in politics as well as medicine.

Illinois Health Director Frank Jirka's attempt to compete with Herman Bundesen's baby books and Bundesen's bid for the Democratic nomination for governor were two of the more visible proofs that the Chicago Board of Health could not escape partisan politics. Whatever the political hijinks, the surgeon general's office attempted to stay out of the fray—at least visibly. For example, when Jirka announced his resignation in the fall of 1937, R. A. Vonderlehr wrote to O. C. Wenger that "it would not be the proper thing for the Public Health Service to attempt to influence from Washington." He went on, however, to offer his "own personal opinion" that "it would be favorably received" in Washington if the governor would request the assignment of an experienced Public Health Service officer to this position.[2]

Two months later, Public Health Service Surgeon David C. Elliott wrote with some consternation to Senior Surgeon K. E. Miller that the governor of Illinois was considering the appointment of Dr. Andrew Cosmos Garvy, a professor of surgery with no public health qualifications (a prerequisite for the job). Could the surgeon general offer a ruling or opinion, Elliott wondered. Miller repeated Vonderlehr's avowal of noninvolvement but added that because the Health Service had jurisdiction over the budgets of the Social Security, the office *did* have the authority to "disapprove that portion of the program which may be affected by such personnel."[3] Although the exact details of the event are unknown, Garvey was not appointed; Dr. A. C. Baxter, already in the state office and named acting director upon Jirka's resignation, moved into the permanent position, apparently with no objections from Washington. Whatever the circumstances, the surgeon general's office was kept carefully apprised of the situation, ready to exercise its prerogatives if necessary but in the name of personnel qualifications rather than politics.

The administration of a federal program through state and city departments of health ran into other political controversies. As with personnel issues, however, the Public Health Service tried to remain on the sidelines. For example, in the spring of 1939, Health Commissioner Bundesen and Board of Health Secretary Louis Schmidt were suspended for about six months when the city of Chicago, fourteen corporations, and forty-three individuals were indicted in an antitrust suit alleging the control of milk prices.[4] The suit eventually failed, and Bundesen and Schmidt were reinstated upon Mayor Edward Kelly's reelection, although Schmidt was not returned to his position as secretary.[5] The affair upset the normal business of the program, however. At one point, Wenger wrote a short-tempered note to Vonderlehr: "I have too much grief with the syphilis problem to worry about this milk affair."[6]

Among the lower ranks of the program as well, relations were occasionally strained. As the summer of 1938 wore on, Reitman commented that many of the WPA employees hired as clerks for the project were being treated with condescension and hostility by other city employees. "The feeling is growing and something ought to be done to change the psychology of the government city hall workers and the WPA," Reitman cautioned. He advised that services, social activities, and training programs be created for the WPA employees that would "unify and solidify" both groups.[7]

Part of the "psychology" undoubtedly grew from the ongoing pressures on the various relief programs, all of which were financially pressed during these years. At one point during the period, Chicago was home to seventy thousand WPA employees and another 225,000 people on relief.[8] As the antagonism between Mayor Kelly and Governor Henry Horner grew, Horner signed or vetoed bills that resulted in less and less state relief to Chicago, placing more and more stress on the city's diminishing funds. In the fall of 1937, WPA allotments to Chicago also fell. Kelly insisted that the city's dole could bear no more, but Springfield remained adamant in its refusal to help.[9]

At that same time, Reitman noted that the growing inability of Chicago's relief programs to accommodate all of its applicants was threatening to place a serious, additional burden on Chicago's health care system. "The Chicago Relief Administration stated that it was going to take thousands of people from the Relief rolls," he reported, "and there is very little evidence that industry will absorb many of the unemployed. The public-spirited citizens who usually gave liberally to charity have gotten out of the habit and think that the Government, State and Municipality ought to take care of all the poor. The M.S.H.C. ought to be prepared to carry a heavier load this winter."[10]

Public attitude toward various relief programs, however, had gone from skeptical to critical. At the start of 1938, the *Tribune* ran a lively series of accounts, replete with photographs, of various reporters, women as well as men, who disguised themselves as hoboes and went in search of help. Reporter Clifford Blackburn, posing as Eddie Brennan of Omaha, recounted his experience in "J. Boyle, Hobo, Finds Chicago Relief a Cinch." Blackburn reported receiving a ridiculously cursory medical examination, including a quick glance for syphilis chancres, from a Dr. G. Kaplan.[11] Two days later, undercover reporter Seymour Korman told readers that his doctor, C. Scruggs, in reference to that news story, jokingly asked, "Any TRIBUNE reporters in this crowd?" before performing his equally perfunctory examination. Korman went so far, in the quick interview that qualified him for relief, as to imply that he was a fugitive from justice but was nevertheless put on the rolls with "no questions asked."[12]

Public debate over the use of welfare money from all levels of government was part of a wider dissatisfaction with political spending that extended to fundraising as well. At one point, this dissatisfaction created a scrutiny that might have tripped up the Chicago Syphilis Control Program. In January 1940, the new state director of public health, A. C. Baxter, asked Public Health Service Regional Consultant C. C. Applewhite whether state employees paid by Social Security money should make the same 2 percent contribution that other state employees made "to the political machine" of Illinois. Applewhite answered that because the Social Security money was deposited in state coffers, it was subject to the same restrictions (or lack thereof) as other state monies. Thus, the federal Hatch bill, which prohibited such political deductions for *federal* employees, did not apply to *state* employees paid by federal money.[13] Applewhite's decision was upheld by Surgeon General Parran's office, which professed no interest in "the question of contributions by State employees to organizations" but wanted only "to assure itself that a satisfactory program of public health [was] being carried on and honest expenditures of funds [were] being made in accordance with the budgets which have been approved by the Public Health Service."[14] Such an answer corresponded with the Health Service's stance of keeping to the sidelines in state and local decisions as much as possible.

But the story was not to end so easily. Two months later, the Hatch bill came up for reapproval in the national legislature, and Republicans tried to extend the prohibition to state employees paid with federal funds. The "two per cent club" of Illinois and similar organizations in many other states were the target of the proposed legislation. Democrats split over the issue, and the debate was followed closely on the front pages of the *Chicago Tribune* during the first two weeks of March.[15] By the end of that time, however, a second story was jockeying for front-page space. An Illinois police officer had filed a complaint charging that he had been forced to contribute 2 percent of his salary to the Democratic party, money that was, he claimed, being used for personal expenses by some party officials. The ensuing indictment froze the fund, which was needed to finance Democratic campaigns in the upcoming spring election. Next, a warden at Joliet filed a complaint that he had been pressured to contribute 20 percent of April's paycheck to make up the campaign chest that had been lost with the indictment. On Sunday, March 10, readers learned of the death—a probable suicide—of F. Lyndon Smith, until recently the manager of the organization and funds of the Illinoisans Club, the keeper of the 2 percent "contributions" from state employees.[16]

What would an investigation reveal? Allegations abounded, and accusations that federally funded employees in Illinois were contributing to the state's Democratic slush fund hit home with the Chicago Syphilis Control Program. In a personal letter to David Elliott, Vonderlehr now commented tersely, "This is not a proper use of these funds, and I trust that you and Doctor Applewhite will be able to obtain information which either confirms or disproves this allegation."[17] Apparently, however, the question of the 2 percent club and employees connected with the program was resolved (or sidetracked), because no further reference to it appears in the federal correspondence of the program.

Relatively free and open use had been made of some federal monies and employees in the past. Roger Biles reports that in Bundesen's gubernatorial race in 1936 numerous complaints were filed about WPA employees being "advised" to contribute to or work for the campaign, charges against which the WPA's director Henry Hopkins took no action.[18] Biles concludes that "the Kelly-Nash organization thrived as a result of its association with the New Deal." Arguing with other analyses that the New Deal destroyed urban political machines, Biles observes, "Clearly, the relationship between Kelly and Roosevelt was both cordial and mutually beneficial. No evidence indicates that Roosevelt set out to de-emphasize the achievements of the local Democrats in favor of his own federal programs. Clearly, the ties between Chicago and Washington grew stronger, not weaker, during the New Deal."[19]

Biles's analysis supports the pattern repeated through various incidents involving the U.S. Public Health Service and the Chicago Syphilis Control Program. A federal stance of noninvolvement was often relatively superficial, with a greater leniency at times permitted if Washington's ends seemed best served by doing so. In the case of the Health Service's first attempt at wide-scale syphilis control, Washington's ends were often served by Chicago's mayor and commissioner of health.

"*A New Federal Racket*"

To the Republican *Tribune*, the New Deal had been a hard pill to swallow. The newspaper had been attacking the federal administration with growing vigor for several months. Its desire to expose abuse of both national and local relief programs—abuse by both the givers and receivers of that relief—followed its traditional opposition to governmental welfare programs. From the earliest days of the Chicago Syphilis Control Program, the *Tribune* had been in the paradoxical position of supporting a campaign funded in large part by a system it opposed politically. In early February 1939, a year before the uproar over the 2 percent clubs, the *Tribune* could not keep the two elements separate any longer. It began covering the movement through the Senate and House of the LaFollette-Bulwinkle bill, which proposed to provide funding for a national antisyphilis campaign. As originally proposed by Wisconsin senator Robert LaFollette, the surgeon general would annually determine appropriation to the states of federal dollars for "the investigation and control of the venereal diseases."[20] The bill underwent several modifications as it passed through the requisite Senate and House committees, eventually passing both houses with a budget of $15 million, paying $3, $5, and finally $7 million a year over three years.[21]

When President Roosevelt signed the bill into law on May 25, 1938, Reitman noted the event with pride in Chicago's accomplishments in all of the areas that the bill targeted. He wrote, "The bill encourages state and local health departments to grants of funds, to wage the war on nine fronts" that were already "in full operation in Chicago," regretting only—but vehemently—that education still did not include teaching chemical or mechanical prophylaxis. A few months earlier, Reitman had copied into his report an earlier version of the bill, which called for "assisting states, counties, health districts, and other political subdivisions of the States in establishing and maintaining adequate measures for the prevention, treatment, and control of venereal diseases."[22] The final version dropped the word *prevention*, an amendment that Reitman immediately caught. He retorted, "Mustn't say the naughty word, mustn't say the naughty word. Christ! what's going to happen at the day of reckoning?"[23]

The *Tribune* also expressed reservations about the bill but for quite different reasons. It had dutifully followed the odyssey of the LaFollette-Bulwinkle bill over the weeks, placing itself strongly in opposition to the measure from the start. One editor charged, "As the father of the national fight against the syphilis plague, THE TRIBUNE regrets to see it turned into a new federal racket." Although the editorial expressed pride in having "favored honest and practical measures for the conservation of natural resources and still does," the editor had "lived to see this policy perverted and pushed to near communistic extremes." He went on to liken the LaFollette bill to the WPA and the corruption with which it felt that program was riddled.[24] Reitman's response was quick and cutting. "The Tribune is voicing the opinion of organized medicine and the private physician," he criticized, "that the outcome of the syphilis drive is toward state medicine, health department control of syphilis cases, and free hospitals and free treatment for all syphilis cases. This editorial is but the beginning of an opposition to the National syphilis drive."[25] Clearly, the *Tribune* was in an awkward position, opposing the source of continued funding for the program whose special advocate it had been.

The *Tribune* was not alone, however, in its disapproval. Chicago's medical community had a long history of opposing reduced or structured payment plans for health care. Both Doctors Schmidt and Bundesen had been targets for the fury of the Chicago Medical Society over the creation of the Public Health Institute and the United Medical Service many years earlier. The Chicago Medical Society also opposed the creation of two more similar clinics in 1935 and 1938.[26] In 1938, Reitman noted additional tensions when he observed that the society, "a special group of physicians or a combination of medical men and business men, including the druggists, might get together and influence the Mayor or the City Council to close the M.S.H.C. and turn all the patients over to the Medical Society, private physicians, or special clinics."[27]

On the national level, organized medicine's fight against "socializ[ed] medical practice" was led by the Chicago-based American Medical Association (AMA) and its journal's editor, Morris Fishbein.[28] It was a battle that many members of the medical profession had been waging for years. In 1920, Congress proposed legislation that required states to create child hygiene or welfare bureaus in order to receive federal money for health programs for children and mothers. It was the first bill of its kind for health care, and physicians were one of the only two groups opposing it (the other was the National Association Opposed to Woman Suffrage). One senator spoke against the bill, quoting a statement from physicians in Massachusetts who said that "control by the individual is democracy, control by the state is autocracy and in other words socialism."[29] Although the Sheppard-Towner Act passed, it continued to excite opposition, and when a continuation of the bill was requested five years later, the AMA was the most vocal group opposing such an allocation, calling it "an imported socialistic scheme unsuited to our form of government."[30] Fifteen years later, in 1935, when mothers and children, along with the elderly, dependent and crippled children, and the unemployed received funding through a new bill, the Social Security bill (which also provided funds for syphilis control), the AMA continued its attack.

The Social Security bill passed, however, and Ann L. Wilson notes that the severity of the Great Depression soon led to a broad acceptance, at least outside the medical community, of the federal government's exercise of such authority.[31] By then, the AMA's ire was directed toward two other perceived threats. In the summer of 1938, some hospitals in Chicago had begun offering medical plans in which a small yearly fee guaranteed twenty-one days of free hospital care for members. At its annual national meeting, the AMA passed a resolution to forbid medical technicians to serve patients with such coverage.[32] As the AMA took this action it was moving toward a showdown with President Roosevelt and the Department of Justice. In February 1938, the President's Interdepartmental Committee had proposed a comprehensive health program that would expand public health services for mothers and children, expand hospital facilities, provide medical services for the "medically needy," guarantee compensation for wages lost during illness, and—its most controversial aspect—create a General Program of Medical Care through special taxes, employer contributions, or both. It was this last point that caused the strongest reaction from the AMA.

Physicians in the association argued strongly on both sides of the proposal, but the trustees of the AMA finally adopted a position opposing the plan. They offered a compromise—to support all points of the program except the General Program of Medical Care. Their compromise was rejected, however, and as a consequence the Justice Department immediately began investigating charges that the District of Columbia's branch of the association was violating antitrust law by obstructing the operations of a prepaid group of Washington's Group Health Association. As contention over the charges grew, the Justice Department subpoenaed the AMA to produce all the documents it had, from as far back as 1930, pertaining to eight low-cost or prepaid health programs around the country—three of them Chicago services and the fourth the Illinois Social Hygiene League, of which Louis Schmidt and other prominent Chicago physicians were leaders.[33] These events eventually culminated in the filing of an antitrust suit against the AMA that dragged on for five years.

Through both circumstance and its history of what by many standards was radical innovation in medical practice, Chicago found itself enmeshed in the test of the power of organized medicine to oppose government's movement toward the nationalization of medical care delivery. Dependent on both the cooperation of private physicians and the Roosevelt administration, the Chicago Syphilis Control Program was inextricably caught up in the conflict. The position of the Chicago Board of Health, led by men who had nearly always been at odds politically and professionally with the Chicago Medical Society, became increasingly awkward. By its participation in the Syphilis Control Program and whole-hearted support of the latest hospital prepayment plans, the Board of Health seemed uncowed by the Chicago Medical Society.[34]

Reitman, however, and perhaps predictably, felt that their position was not strong enough. "In Illinois the state health department is influenced by the state medical society to a dangerous degree," he began, drawing a picture of a society that had relentlessly opposed or tried to manipulate all of the city's public health efforts. He cited the Chicago Medical Society's initial opposition to medical relief in Chicago: "Now," he accused, "they have control of it." The society had "never had the health and the benefit of the people at heart." The only benefits of interest to them, Reitman accused, were their own, and he turned directly to the Chicago Syphilis Control Program for his proof. "As long as the gonorrhea and syphilis section of the Chicago board of health was an insignificant division the medical society paid little attention to it. Now when it is a powerful and growing clinic with 2000 patients a day and when it has a budget of $2,000,000 a year the medical societies are reaching for it to take control."[35]

However accurate Reitman's analysis, the opposition of organized medicine to nearly any form of nationalized health care was evident. Coupled with growing public opposition to other federally funded social services, this additional controversy further imperiled the popularity of the Chicago Syphilis Control Program. Maintaining public enthusiasm for a program whose medical outcomes seemed already to be reaching a plateau would be difficult in the best of times. Now, its sources of funding were under increased public criticism, and its success was partially dependent on an increasingly hostile medical community—and there were even further complications.

"The Decent People Are Opposed to It"

With the exception of State Assistant Epidemiologist Dr. A. J. Levy's public statement that "every youth over seventeen should be taught to apply prophylactic measures," similar recommendations from program officials had been made only internally.[36] Even Reitman's public advocacy of prophylaxis was kept carefully separate from his work with the program—or was *attempted* to be kept separate: "But periodically we leave the role of investigator, statistician and become the Voice of the Prophet, crying in the Big City, Prepare ye the way for Prophylaxis," Reitman confessed.[37]

His prophesies, however, seldom made it into any print other than his own internal reports, so it was a red-letter day for Reitman in July 1938, when Louis Schmidt, who was the secretary of the Board of Health, a prominent member of the Illinois Social Hygiene League, and the long-standing chair of the mayor's prestigious antivice Committee of Fifteen, was interviewed at length about the necessity of teaching and developing means of preventing venereal disease. He was quoted in the *Tribune:* "Education of the public in methods to prevent infection is necessary if the city's campaign against syphilis and gonorrhea is to be fully successful. . . . Venereal disease spreads faster than our clinics and physicians can cure it." When the reporter observed that there was no vaccine to prevent syphilis, Schmidt replied, "There are, however, methods of prophylaxis—mechanical and chemical means which, when secured from reliable sources and properly applied before or after intercourse, prevent both gonorrhea and syphilis. . . . But we have still to conquer social resistance to prophylaxis, the most practical means of prevention."[38] Although the article did not go on to say what those means were, Schmidt's statements were remarkable nevertheless, seeming to indicate a clear shift in the public posture of the Board of Health.

Schmidt's statement was the very one that Reitman had been waiting months to hear. "The Lord must love the Chicago Board of Health," he crowed, "especially the syphilis and gonorrhea section. Because it is continually increasing in usefulness and is growing in wisdom, and it's pioneering so many important phases of syphilis control."[39] Reitman's optimism was short-lived, however, because the rhetoric soon returned to normal, and his attempts to talk about prevention were once again ignored or rebuffed. Three months later, he reported "haunting" a meeting of the Safety Congress, where Dr. A. E. Russell of the Syphilis in Industry Program presented a paper on industry's efforts in the fight against venereal disease. Refusing Reitman's request to include information about prophylaxis in his talk, he told Reitman, "Headquarters [Public Health Service officials] do not look upon venereal prophylaxis favorably. . . . The clergy, the woman's clubs and the decent people are opposed to it."[40]

In the discussion following Russell's presentation, Reitman raised his hand. "This is the representative for the Chicago Society for the Prevention of Venereal Disease," he reported himself as saying, then, "What would you say if we told you that the machinery shops in a big industry was killing and maiming thousands and then said, we have a splendid hospital and very fine surgeons to take care of the dying and the wounded. Syphilis is a preventable disease. No one need ever catch it. Wash the parts with soap and etc. Put up a placard in all of your industries 'It pays to be decent. Be loyal to your wife. Extra-marital contacts are [dangerous]. If you indulge use prophylactics.' "

Russell's first response to Reitman echoes the Chicagoan's conversation with Dr. Vonderlehr ten months earlier. Where Reitman's report of that earlier exchange, though, was fired with righteous anger, his later response sounds despondent and worn. He concluded his report of the incident without his usual diatribe or rallying cry. Instead, "There was no applause," Reitman noted. "Russell never [mentioned] it in closing the discussion."[41] This account is significant for more than Reitman's shift in tone. Two copies of this page of Venereal Disease Control Report Number 57 are included in Reitman's papers. Both texts are identical with one exception: Typed across the top of the second version, underlined, is the heading "*The Last Report that Ben Reitman Made Before He Was Fired.*"

Although Reitman may have been out, he wasn't down. By December 29, he was writing to his old friend and mentor William A. Evans, now retired to Mississippi, about a meeting of the syphilis division of the Board of Health that he had attended "uninvited and unofficially." About his recent dismissal, he wrote, "Parran, of the U S P H S denies that he had me fired but Wenger boasted of it. But none of the health agencies including the W.P.A. in Chicago will take me on, knowing that Parran is opposed to my prophylaxis activity. Nevertheless we have continued our prophylaxis work stronger than ever before. These ten weeks being without a job have been very precious. We have been able to do work which we could not have done had we been officially attached."

Reitman's account of his dismissal, however, was strongly disavowed by the U.S. Public Health Service. When he complained to Assistant Surgeon General R. A. Vonderlehr that he had been fired at the direction of the

Health Service, Vonderlehr sent a quick refutation of the charge, writing that "the employment of personnel in connection with the venereal disease control activities in States and local communities is a function of the State and local health officers."[42] Reitman was basing his accusation on the reason given him by Herman Bundesen, who reportedly said, "There are two members of the Reitman family on the payroll and the U.S. Public Health Service won't stand for it."[43] Reitman's wife, Medina, worked in one of the city's medical laboratories, a situation that had been known and unchallenged for some time; what is noteworthy is that Vonderlehr's defense was to defer the authority of the Public Health Service to that of state and local departments, the tactic used on most other matters of political delicacy.

Despite his outburst, however, Reitman could not afford to alienate the ear he felt he had in Washington. Since 1937, he had written regularly to Vonderlehr, often sending copies of the reports he was writing for his bosses in Chicago.[44] Vonderlehr responded to many of Reitman's reports personally in the first year of the program, but by the late summer of 1938 he frequently deputized other officials to answer for him. Letters politely expressed thanks to Reitman for his continued interest in the Chicago Syphilis Control Program but seldom went on to engage in the discussions that Reitman initiated. This cordial tone was broken only once—in this instance of official response to Reitman's angry accusations about his termination.

Faced with Vonderlehr's shortness, Reitman quickly sent an indirect apology, mailing to Parran copies of correspondence with Nels Anderson, an old acquaintance of Reitman and a noted sociologist currently working on the Labor Relations Board of the WPA. Anderson had taken Reitman to task for his anger, writing, "Normally I do not quarrel with direct action in dealing with social problems, but the crusading position you take with reference to public information about syphilis is one that no public agency here or abroad would be able to support. You presume to take liberties with the mores, and you become irritated when you encounter public officials who may be required, by oath of office and accepted practice, to disagree with you." Anderson contrasted the freedom of private physicians to advocate or prescribe prophylaxis with the limitations that public officials faced. He believed it unthinkable, "if you have any respect for democracy," for public health officers to use their delegated power to dictate other people's behavior. At the same time, Anderson paid gentle tribute to those very inflammatory qualities that gave Reitman his special appeal: "What you need is a good course in public administration. I am afraid that if you took such a course and it soaked in the result would be disastrous. Your best friends would no longer recognize you."[45] This time Parran deputized Vonderlehr to respond to Reitman. He wrote that Parran "was interested in the letter sent you by Mr. Anderson. As this gentleman states, tolerance is an admirable quality for one to possess at the present time."[46]

The loss of his position in Chicago's work against syphilis was a blow to Reitman, and his anger about the lack of a reasonable program of prevention was matched by regret at being shut off from work that was still of vital interest to him. He was aware of the force of his emotions and that they were sometimes irrational. Anderson's argument about the most democratic approach to presenting his views on prevention was probably one of the most useful ones to use with Reitman, whose analysis of syphilis control was always political. Thus, he passed the months following his termination, his anger frequently competing with his desire to continue to be a part of an effort that he saw, with all its limitations, as sharing his desire to rid the world of syphilis.

In fact, Reitman did hold one more job with the Board of Health. He was a smallpox vaccinator in the transient district for the winter months of 1939. The job was procured for him through political connections of his brother Lew, whose General Outdoor Advertizing Company put up billboards for many of Chicago's politicians.[47] Perhaps not surprisingly, the duration of that job coincided with the short period that Health Commissioner Bundesen was on leave of absence without pay during the grand jury investigation of Chicago's milk business.[48] Reitman's dismissal from his last city job came close on the heels of Bundesen's reinstatement.

"For Every Jack"

If Louis Schmidt's words calling for prophylaxis had thrilled Reitman the summer before he was fired for advocating similar measures, he must have felt a bittersweet vindication when he read the annual report of the Chicago Syphilis Control Program for 1939–40. Although previous annual reports had also insisted on the importance of prophylaxis, this one finally offered more than a principled argument. In the report's list of seven "Plans for 1940–41," the seventh item read: "To distribute prophylactic information to all males having blood tests." (An obvious question here is why only men were to receive the packets; the answer will soon become apparent.) The closing summary of the report indicated the program's officers' awareness of the current program's limitations. It read:

> We believe that the next step should be widespread dissemination of prophylactic information. These measures have proven highly successful in the military and naval services and, with the National Preparedness Program in full swing, the time is ripe for the introduction of similar measures to the civilian population.
>
> It is well known to every health officer and physician that there *are* people who continue to expose themselves to infection, despite repeated warnings and bad examples. If we cannot prevent people from exposing themselves, we must teach them to protect themselves.
>
> The introduction of such measures is essential to the ultimate success of the program.[49]

In three years, the officers of the program had carried its testing and treatment phases as far as possible. Laws now required premarital and prenatal testing for syphilis, numerous industries and colleges routinely tested all new employees and students, the Board of Health and the Chicago Medical Society offered regular classes on venereal disease to physicians, and more private physicians than ever tested patients and took advantage of free drugs from the Board of Health. With all these programs in place, fewer people were being tested than the year before, and fewer cases of syphilis were being found.

Although the drop in positives was due in part to an overall increase over the years in the number of new cases discovered and brought under treatment, the authors of the Annual Report noted that the decline in the number of tests given that year could also be accounted for by the closing of the dragnet stations, a reduction in staff from the WPA, and a general waning of the novelty of the program. On this last count, the report read, "Last, but not least, is the fact that the first blush of enthusiasm has worn off. . . . There is now ahead the more difficult task of convincing those groups and individuals not yet examined of the importance and value of this service."[50] It was difficult, especially among poor patients, to keep people in treatment for the full course of chemotherapy, and some would simply refuse testing—or refuse to refrain from acts that could produce infection.[51] By officially stating that people would not give up sex to prevent venereal infection, the Chicago Syphilis Control Program seemed, finally, to be saying in print what Reitman had been urging them to recognize or admit since the start of the program. The annual report was labeled "confidential," but if the recommendations were to be heeded, a public reversal of the U.S. Public Health Service's position on venereal prophylaxis would have to occur.

From Reitman's standpoint, advocating the prevention of venereal disease by other means besides sexual abstinence was simple common sense. There was, however, another impetus for federal officials to heed this advice—war. No reader of the *Tribune* during the late 1930s could escape the steady and seemingly unstoppable acceleration of fighting in both Europe and the Far East. The 1939–40 Annual Report of the Chicago Syphilis Control Program reflected this national concern. The reference to the National Preparedness Program in the concluding paragraphs of the report was no coincidence. Within the annual report, a "Special Report" observed that if "the same positive rates [for syphilis] prevail in other large cities of our country, we will find enough infected men to form an army corps." "To make matters worse," the report continued, "our experience here in Chicago has shown that approximately 50 percent of the cases uncovered and brought under medical care in the age group *21–31* years are classified as *early syphilis*. . . . In other words, with an estimated load of *11,054* cases of syphilis in Chicago among men in draft age we have over 5,500 early infectious cases. Think of it! 5,500 individual little fires, just waiting for a chance to spread."[52]

The "spread" of these "little fires," however, was seen to threaten not the women who would subsequently be infected but the other soldiers who these women would, in turn, infect. "In a civilian population this is bad enough," the authors of the report admitted, "but in an army encampment, where the sexual activities of the men are more or less confined to a limited number of women, it is disastrous. When these infected soldiers get to the camps without being detected they infect the women around the camp, who, in turn infect other soldiers formerly free from the disease, and thus a vicious cycle is established and maintained."[53] That the thrust of this concern was to have a ready army was made clear in the next statement, printed in capital letters and underlined a word at a time: "ALL THIS SHOWS THE NEED FOR AN INTENSIVE DRIVE TO UNCOVER SYPHILIS IN MEN OF DRAFT AGE AT THE TIME OF REGISTRATION."[54] By testing men at registration rather than enlistment, infected men could immediately be placed under treatment by civilian physicians at civilian expense and thus "be ready for service" when they actually enlisted. "Once a man is enlisted, sent to an Army camp, and then found to have syphilis, he must either be treated at the camp—where he is a continuous hazard to the uninfected soldiers of the command, as shown above—or be returned to the point of enlistment for treatment, *all at government expense*."[55] Moreover, a soldier treated in camp for syphilis automatically became eligible for "early pension, hospitalization and other benefits for himself and family" for all subsequent health problems related to the disease.

"Now that's not all," the report continued, "For every Jack there is a Jill and the infected Jill, even if she is not enlisted, is even a greater hazard than Jack to the military forces. Just as the young man inducted into the Service is removed from the restraining influences of his home environment, the young girl is swept off her feet by the glamour of her brother and boy friends in uniform, and so ethical and moral standards become less defined. This always means more syphilis."[56] The statement invokes two traditional images: the portrait of women easily swayed by the romantic appeal of a man in uniform, and the menace of a syphilitic Jill. That Jill was considered a greater danger to young American soldiers than a syphilitic Jack was to her, however, led to Wenger's recommendation that "all women uncovered with venereal disease should be immediately, from this moment on, given treatment under the strictest quarantine regulation possible. We should not wait as we did in the last war."[57] Such statements help explain why the more general recommendations of the program advised giving prophylactic packages only to men who were tested. Even though this recommendation could also reflect the unavailability of a form of "medical" prophylaxis (as opposed to soap and water) for women and the assumption that condoms would only be carried by men, it is nevertheless clear that the maintenance of a healthy fighting army was of greater concern than the health of women. Moreover, although Reitman, in previous years, had challenged the double standard that would isolate or quarantine only infected prostitutes, we hear in this last recommendation an even broader application of a sexual double standard that would quarantine women (but not men) engaging even in "legal"—that is, free—sex.

In many ways, the report of the Chicago Syphilis Control Program demonstrates an attitude not substantially advanced from the attitudes toward syphilis during World War I and the Spanish-American War. Although by 1940 sexual abstinence among men had been abandoned as an unrealistic goal (a major step in the direction of being realistic), the need to protect "our boys" from disease-carrying temptresses was still a prevailing image. Sex was once again being painted as dangerous and sinister, an inherent part of the chaos and emotional vulnerability of young men facing death on the battlefield.

By July 1940, many of the goals of the Chicago Syphilis Control Program had been achieved—at least as far as possible within its current structure and policies. With this proclaimed success, and with the national focus of concern about syphilis shifting to a wartime stage, much of the civilian work of the program was ended: Recording and reporting systems were in place; clinic procedures were established and could now be handled in the new and enlarged city facilities rather than in special clinics and field stations; the large number of WPA and other relief workers could be reduced as a result of both natural reductions in the number of staff needed, waning federal and local relief funds, and increasing hostility toward New Deal hirees; and no new, unsuspected pockets of infection were expected. Public health programs had begun to incorporate the new, fuller range of services for venereal disease into their existing structure. Journalistic attention also shifted to new arenas. The war on syphilis gave way to the wars against Hitler and the New Deal. By the last months of 1938, stories on the Chicago Syphilis Control Program all but disappeared from the pages of the *Tribune.* Following the November elections the newspaper's headlines proclaimed, "New Deal Wrecked by G.O.P.; Revolt against Roosevelt Sweeps Nation."[58]

Ben Reitman managed to maintain a modest medical practice, but he was supported increasingly by his busy wife, Medina Oliver. He continued to write old bosses for new jobs and to offer them his unique commentary on the city's and nation's efforts against syphilis, but he found most of his joy in the personal dimensions of his work. In 1941, he wrote to Reuben Kahn, "Any way, it's a good world, and an alcoholic, syphilitic woman who was worried about being pregnant bought a new battery for my car."[59] The world—and Reitman's car—moved on.

CHAPTER ELEVEN

Aftermath

Because this story is shaped by the various voices that spoke of or for the Chicago Syphilis Control Program, it can continue only as long as those voices were heard. Ben Reitman's daily reports to Lawrence Linck ended in November 1938, but he continued to write letters to his old bosses in Chicago and Washington until his death in November 1942. Reitman's passing was noted by articles in several local newspapers, where he was remembered for his anarchist past, the con men and hoboes who attended his funeral, and the party for the homeless for which he provided in his will. An "eccentric apostle of many isms," the *Chicago Tribune* described him. His work with either syphilis in general or the program in particular went unnoted.[1] But the nation had other concerns by 1942; as early as 1939, the *Tribune*'s vigilant coverage of the program had virtually ended. As the national syphilis control effort moved into a wartime venue, new organizations—and new functions for existing organizations—caused the work in Chicago to become more diffuse, until a whole new cadre of city, county, state, and national programs for syphilis control was created.

Any assessment of the program's achievements will lead to a range of conclusions. On the one hand, the continued existence of syphilis offers unassailable proof that the program failed in its pledge to defeat the disease. On the other hand, because the program succeeded in bringing more cases of syphilis under treatment in Chicago than ever before, a clear pragmatic triumph could be said to justify the hyperbolic rhetoric that surrounded the campaign. Between these two extremes, however, lies an analysis that recognizes advances but pauses to examine their limitations and consider, once more, the interplay of structures and motives that would have confounded any undertaking as ambitious as the Chicago Syphilis Control Program.

Medical Achievements

In October 1939, Lawrence Linck wrote to Ben Reitman, "I am enjoying the experience of working with people who by their actions as well as by their words appear to be working toward a common goal—that experience is a little unique for me after the past two years. . . . Your constant uncompromising outlook has accordingly been a source of considerable satisfaction to me."[2] Linck's words support the supposition that he, like Reitman, was let go, for whatever other reasons, for not always tempering his words and opinions. However disparate the goals of various leaders of the Chicago Syphilis Control Program might have been, their cumulative effort and achievements were substantial. So, for their time, were their resources. The program's budget grew steadily, from $104,149 in fiscal year 1936; $290,182 in 1937; $442,188 in 1938; to $828,775 in fiscal year 1940.[3] Although Chicago continued to support the program, the city's contribution steadily decreased from 81 percent of the budget in fiscal year 1936, to 43 percent in 1937, then 36 percent in 1938, and 31 percent in 1940. Although the investment did not ultimately eradicate syphilis, it did accomplish some remarkable changes.

In the first four months of 1938, a year after the national kickoff of the antisyphilis drive but before the formal opening of Chicago's dragnet stations, three times more syphilis tests were being performed than had been during the same period one year earlier: 88,714 tests compared with 28,047 from January through April 1937.[4] From July 1938 through June 1939, 11 percent of Chicago's population—371,693 people—had been tested.[5] A year later, another 10 percent of the population—326,933 people—had received diagnostic tests (rather than tests on people already in treatment).[6] Program officials, disappointed that the percentage had fallen, attributed the decrease to several causes: the closing of the dragnet stations (which, the program failed to note, had been closed *because* they were doing poorly), the loss of some WPA staff, and the fading of the "first blush of enthusiasm" for the program.[7]

Yet, although the number of people tested dropped by 1.4 percent, the prevalence of infection fell even more dramatically. Where 7 percent of those people tested in 1938–39 had been found to be infected, the overall rate the next year was only 4.9 percent.[8] One conclusion that could be drawn from these numbers—and the one presented most often by spokespersons for the Chicago Syphilis Control Program—was that the first year or so of the program had brought under treatment many cases of "old" syphilis. As these people were discovered and entered into the records, only newer cases remained to be found. Programs were also casting an ever-broadening net, particularly through laws for premarital and prenatal testing, the voluntary testing of schoolchildren, and the surge of voluntary (and growing mandatory, preemployment) testing in many businesses and companies. The diminishing amount of preexisting syphilis as well as the growth of mass testing would naturally result in a "drop" in rates.

Still, the number of people brought into treatment and the extent of Chicago's involvement were sizable, and the diversity of the program's clients grew between 1936 and 1940. Where only two hundred of Chicago's six thousand private physicians had "sent occasional serologic specimens to the Municipal Laboratory" in 1937, 4,732 did so in fiscal year 1939–40.[9] Even so, the bulk of private care was being handled by fewer than one-third of the city's private physicians.

Although program officials praised the increase in participation, they still bemoaned their inability to persuade most private physicians to cooperate. Legislated testing programs, now fully in place, performed 12,104 prenatal and 32,127 premarital examinations, uncovering infection rates of 3.6 percent and 1.7 percent (down 1 percent from the previous year), respectively. The school testing program discovered syphilis in approximately .7 percent of students. The city's Ehrlich Clinic, an adjunct of the Chicago Maternity Center, treated pregnant women who had syphilis and were from the poor neighborhood surrounding the facility, where "practically all colors, race, and creeds [were] represented."[10] The new Gonorrhea Unit of the Chicago Municipal Social Hygiene Clinic reported giving ninety-six thousand treatments in fiscal year 1940. A special nutrition program had been created when practitioners noted some of the depleting side effects of the chemotherapy as well as the malnutrition of many of the poorer patients. Another temporary program was also established to provide clothes and shoes to particularly needy patients.[11]

Clearly, there was less untreated syphilis by July 1940 than there had been even two years earlier. Public consciousness had been raised, with the result that more people were seeking treatment and more physicians were participating in the city's programs. Large, previously unattended segments of Chicago's poor men and women were receiving medical treatment for syphilis and gonorrhea, gaining access to medicine and services that had heretofore been unaffordable or inconvenient. Thus, in terms of identifying syphilis and commencing treatment for many people who would not have otherwise learned of their condition, the Chicago Syphilis Control Program was a success.

Long-term Limitations

The broader success of the program, however, is more difficult to assess. Its goals were twofold: to cure all existing cases of syphilis and to stop further infections from occurring. The first goal required public health officials to find, bring to treatment, and *keep* in treatment all citizens infected with syphilis. The second goal required that all citizens—infected or not—become knowledgeable about how syphilis is spread and take personal responsibility for not infecting anyone else. Although achievement of the first goal was abetted by some legal force, it nevertheless depended in large part on individual motivation to seek testing or treatment. Substantial funding and an intensive publicity campaign aided motivation, but the program still did not bring in 100 percent of the population for testing or keep all those who were infected in treatment. The achievement of the second goal, however, required an even greater commitment to voluntary testing and moral prophylaxis. That goal addressed attitudes and behaviors over a longer range of time than the goal for one-time, citywide testing and subsequent treatment. Could it be achieved if the importance of syphilis control was not kept prominently in the public eye?

The preceding chapters have detailed some of the problems inherent in the goal of curing all existing cases. For one thing, without mandatory, repeated testing there can be no guarantee that all cases of syphilis will be detected. For another, laws requiring tests for only special groups can lead to fears of unequal treatment and, consequently, evasion of the law. Moreover, selective testing, such as that practiced in the court system, can also work against open-mindedness toward syphilis. Equally important, bringing people into treatment does not guarantee cure. In 1939, not only was absolute medical "proof" of cure still arbitrary, but the protracted treatment itself also worked against sustaining long-term care. The program reported that following "delinquent" cases was its "biggest problem, amounting to 48 percent of the cases of syphilis it discovered."[12] The next year, the Chicago Board of Health again found that a substantial number of infected people did not complete treatment. The annual report of the program stated, "The most unsatisfactory result in this report is the large number of cases from public blood testing stations and other non-medical sources which have been dropped for such reasons as moved, unable to locate, etc."[13]

Officials of the Chicago Syphilis Control Program saw a connection between poor attendance for treatment and a patient's socioeconomic status, noting the obstacles to full treatment caused by joblessness, tenuous living situations, inadequate sanitary facilities, and transportation problems. They blamed cuts in city and national relief programs that made consistent, sensitive treatment difficult to sustain on such a wide scale.[14] Although such an assessment was not as radical as Reitman's accusations of bureaucracy's insensitivity toward poor or "marginal" people, the program did acknowledge that forces outside the sphere of public health have consequences for public health programs. Nevertheless, the amelioration of the social and economic miseries of Chicago's poor was beyond the ministrations of any one board or program. A program that cannot permanently change the economic situations or opportunities of the people it serves will probably never be able to make permanent changes in their health as a group or class.

The program seemed to combine compassion with pragmatism where it could, as with, for example, its nutrition and clothing programs. Pragmatic, too, it could be argued, was the policy of initially giving only enough drugs to render a person noninfectious, a strategy seemingly designed to protect the public first, a utilitarianism at least superficially reflecting the mandate of public health programs. Still, the availability of treatment to Chicago's poor, African-American community was unprecedented, and, to be certain, many people benefited from it.

Of continued concern is the role that class or racial bias might have contributed both to the decision to stop short of cure even initially and the vigorousness of the effort to keep poor, predominately African-American patients in treatment. Although the reports from 1939 and 1940 showed dramatic decreases in the incidence of syphilis found in both whites and African Americans, "The decrease," announced the Annual Report of July 1939, was "principally in the negro group. . . . It is in this group that we have treated the largest numbers of the population and uncovered the largest numbers of cases." Still, the rate of infection reported for African Americans remained substantially higher—about five times higher—than that for whites.[15]

To what extent was this difference the inevitable result of unremediable circumstance, or to what extent might it have been further reduced by educational approaches, medical services, social inducements, or even professional behaviors specifically tailored to the needs and fears of this rightfully cautious group? For example, although the Public Health Service had a history of working with leaders and spokespersons from racial minorities, few people of color played a visible role in the program. Even within the African-American community there is no clear indication of visible support for the program; instead, federal documents and Reitman's reports reflect the suspicions that many harbored about its methods and motives. Although officials were sensitive to problems that racially loaded statistics created, their decision to avoid discussion rather than seek solutions to this tension ultimately did little to resolve matters. Ben Reitman's occasional comments about the manner of the staff at the program's various clinics show his awareness of the importance of treating poor patients with respect and dignity, and his colorful accounts of his conversations with citizens around the neighborhood dragnet stations were a constant, covert recommendation that officials speak to people in a familiar language and address the concerns that were important to them. Similarly, his repeated recommendations to create full-service health stations or expand the dragnets to offer additional services indicate his awareness that syphilis control is inseparable from health promotion in general.

Of particular note is an unembossed statement by O. C. Wenger in his annual report for fiscal year 1940, in which he credits the decrease of syphilis among African Americans to the "uncover and treat" approach of targeted mass testing while attributing the decrease among the white population to the "increased use of prophylactic measures."[16] This is a curious statement to make, given the surgeon general's proscription against either teaching or advocating prophylaxis. (I am taking Wenger's choice of the word *use* here to indicate the application of prophylactic devices or treatments rather than the *practice* of moral prophylaxis.)

Wenger's statement evokes other historical studies that have remarked upon the upper classes' access to the means of birth control even during times of vocal medical, legal, and moral disapproval of contraception.[17] The contrast he notes makes clear that differential treatment for, or access to, information about syphilis had definite racial overtones. The accumulation of these evidences of Chicago's—and Washington's—reluctance, and at times definite refusal, to grapple with the various issues of race in syphilis control makes it difficult to conclude that the program's decision to place noninfectiousness above cure as its goal rested entirely on pragmatics.

The program's second goal—to stop further infection—was even more daunting than the first, its achievement even harder to assure. It rested, ultimately, on the ability of an awakened public to remain awake to and active in its efforts against the spread of syphilis. Knowledge and participation would be necessary, even beyond the close of the public campaign itself. To this end, the program launched its educational program, although, following the directions of Surgeon General Thomas Parran, most of the public's education was focused on moral prophylaxis or the horrors and devastation of syphilis. In 1938–39, for example, the Chicago Board of Health gave 682 lectures to 56,982 people. It distributed 146,730 pamphlets about syphilis, placed 441 posters in public places, loaned exhibits to 13 groups, and showed movies at 36 meetings. The WPA distributed one million pieces of literature and put advertisements in street cars. WPA workers lectured public school teachers and visited private physicians to explain the various services of the Chicago Syphilis Control Program. Both the Chicago Board of Health and the Chicago Medical Society offered courses, the latter as postgraduate courses from its own venereal disease commission.[18] In the next year, 647 lectures to teachers, workers, and civic club members netted more than 45,000 specimens.[19]

As the program strove for *syphilis* to become the proverbial household word, however, it often called up images of the horrors of lives—and families—ruined by the disease. Although this approach was credited with uncovering many infections among poorer, less well-educated citizens, such an image would, it seems, be of limited benefit in breaking the stigma that surrounded syphilis, especially because many of the testing policies seemed to reinforce its association with the poor, the prostitute, and others who were socially disadvantaged. Much of the language and presumptions of the program, moreover, suggested a condescending paternalism that implied that these groups were unwilling or unable to understand more balanced discussions about syphilis. Wenger's distinction between increased use of prophylactics among wealthier whites and the "uncover and treat" approach used with African Americans also suggests an assumption that the mechanics of prophylaxis were too complex for African Americans to learn.

Finally, emphasizing *moral*—and labeling it as such—over any other kind of prophylaxis for venereal disease still connected syphilis with "immoral"—that is, sexual—behavior. Although Parran's "war" against syphilis may now sound like a "holy" (or holier-than-thou) war, most of the officials of the Chicago Syphilis Control Program realized early on that testing and education alone would not totally cure or prevent the disease. Throughout the months of the program's activities, a steady, underground chorus continued to call for permission to teach all possible methods of prevention. Moral prophylaxis, the only endorsed means of preventing syphilis, proved to be a major impediment to program's ability to achieve all of the goals that the U.S. Public Health Service had set for it. The single greatest failure in the national control effort of the late thirties was Surgeon General Parran's refusal to permit public acknowledgment of people's sexual behaviors or to realize that sexual behavior would continue despite serious risk. Clearly, some learned prophylaxis, whether by condom or soap and water, despite silence on the part of the Health Service or the Chicago Board of Health. This very silence, however, loudly calls into question the repeated statements that the government's war on syphilis was scientific and broad-minded.

International Limitations

The U.S. government's foray into syphilis control both followed and foreshadowed similar Canadian programs whose history has been told by Jay Cassel in *The Secret Plague: Venereal Disease in Canada 1838–1939*, which offers some useful insights for assessing the achievements of the Public Health Service and the Chicago Syphilis Control Program.[20] Canada's first national effort took place immediately after World War I, when the shock at the extent of syphilis among the military and the enthusiastic response to limited programs of treatment spurred officials to undertake a civilian campaign. At this point in U.S. history, it will be remembered, the relatively straightforward information about gonorrhea and syphilis used for soldiers during the war was deliberately silenced by such organizations as the American Social Hygiene Association. Also significantly different was Canada's medical profession (more specifically, "reform-minded" physicians), which urged government to take action because physicians felt the problem was too extensive to be handled locally. This united attack on venereal diseases was the country's first attempt at any nationalized medical program.[21]

The program, called the VD grant, had four arms that were similar to the aims of the Parran's later program: treatment, social service, law enforcement, and statistics. Treatment was always considered of primary importance and was available from both public and private sources. Although more than half of the country's known cases of syphilis were under treatment by private physicians in 1917, Canadians realized that poverty often interfered with the detection and treatment of the disease, so the Dominion provided free drugs for testing and treatment to both public and private health providers, with private patients paying only physicians' fees.

Canada's system was essentially the same as the one that would be developed in the United States twenty years later. "Social service" was defined primarily as following patients under treatment and tracing their sexual contacts. A nurse was assigned to each clinic for this task, which, as with the Chicago program, was perennially underfunded and overworked. Some provinces did not like this service, viewing it as "an exercise in medical policing."[22] Law enforcement, which centered on examining all prostitutes and people under detention, was generally considered unsatisfactory because many magistrates were not willing to prosecute people for having an untreated disease. During the spring of 1922, there was a brief flurry of debate over the prospect of a government-sponsored prophylactic program that would set up "early treatment" stations for men only in red-light areas of cities. At these stations, men would be able to immediately treat their genitals with solutions of permanganate or other chemicals to prevent infection from recent possible exposure to gonorrhea or syphilis. The proposal, however, was defeated, as was a proposal to distribute prophylactic kits to men who visited houses of prostitution.[23] It coincided with the similar experiment, also quickly discouraged, in Chicago.

Generally though, in the first two years of its operation, Canada's endeavor flourished, with enthusiastic reports from many of the provinces. A film about syphilis, *The End of the Road*, played across the Dominion, drawing large crowds and earnest discussion. Emmaline Pankhurst, the noted British suffragist, crossed the ocean following the suffrage victory in England and became a regular lecturer about the importance of sex education for both men and women. People should take responsibility for their own sexuality, was the message in the education program. Responsibility, however, as in the United States two decades later, was defined as containing one's sexual impulses until after marriage.[24] Government officials stifled information about other forms of prevention. Closely tied to official reticence about condoms, Cassel notes, was broad public opposition to discussions of birth control, with which the condom was connected.[25] This position reflected the "social preconceptions, particularly those of the urban middle class from which most of the leadership was drawn."[26]

Prominent among these preconceptions were traditional ideas—echoed in rhetoric in the United States as well—about male and female sex roles. These beliefs and "the insistence that any form of extramarital sex involved a serious risk of exposure to VD and therefore should be avoided" created a campaign that criticized a "life style" as much as it addressed a disease.

Fiscal troubles in the Dominion government in 1924, and success in achieving the first balanced budget since 1913, led to heavy cuts in some programs. Money for the VD grant was steadily whittled away throughout the rest of the decade, and by 1931 no VD grants were made from the Dominion government to the provinces. The Division of Venereal Disease Control itself was eliminated in 1934. Most provinces kept their clinics but concentrated their work primarily on diagnosis and treatment. A brief revival of activity occurred in 1936, a rebirth that Cassel suggests may have been inspired by Parran's initiatives in the United States. Enthusiasm spread once again, but efforts stayed at the provincial level this time, with varied results across the country. Cassel notes problems with a lack of useful statistics, lack of reporting from private physicians, lack of routine standard recordkeeping among clinics, and the inability to assess the extent and effects of tertiary syphilis.[27]

Despite these problems, "the worst ravages of VD were greatly diminished" as a result of the medical achievements of these two campaigns.[28] Cassel is less sanguine, however, about the accomplishments of the two educational programs that accompanied the medical efforts. The mid-thirties saw an attempt to update *The End of the Road* (with an altered plot and the addition of sound) and repeat the popular lecture circuit, but education about syphilis and gonorrhea in 1935 continued to embody the same values as were present during the 1920s. The same problems inherent in the first campaign were present in the second one. Canada's overall efforts toward syphilis control were of only limited success because they fostered (although Cassel does not use the term) syphilophobia: "VD continued to be associated with wayward conduct, irresponsibility, and infidelity, and with infertility and death. In this way the education program made venereal disease public only to drive it back into secrecy."[29]

Although the early support of a large segment of Canada's medical profession and Canada's efforts to move swiftly to capitalize on the medical and attitudinal advances made during World War I mark two significant differences in the two nations' programs to eliminate syphilis, other factors outweighed these apparent advantages. Moreover, they were problems that proved crucial to sustaining any advances made in the Chicago Syphilis Control Program as well. First, an unrelenting federal program was essential—unrelenting in terms of funding, standardization, and oversight. Such diligence required resources that neither country could or would maintain indefinitely, especially when faced with other budgetary crises. Next, whatever support was available from organized medicine, individual physicians needed to play a visible role in promoting, supporting, and complying with the goals of the program, a compliance and support that were always only partial and reluctant in the United States and diminishingly present in Canada. Laws that prescribed—and uniformly enforced—people's behavior toward testing and treatment were also probably necessary, "probably" because such measures would, ideally, not be required had these programs succeeded in their other goal—galvanizing public attention and recruiting unanimous participation. Again, both countries never fully succeeded in either area, with syphilis continuing to make many citizens cautious or fearful, laws often exacerbating those fears and prejudices, and law enforcers often inconsistent in their exercise of the law.

It is significant, however, to see such similarity of goals and approaches in both Canada and the United States, not so much as reassurance that Chicago's program was not necessarily flawed by its own unique actors and circumstances but more to realize the pervasiveness of ideas and values that undergirded these and other similar efforts around the world at this same time. For example, Canada's position toward moral prophylaxis was in accord with official stances in not only the United States but also in Britain, France, Belgium, Germany, Switzerland, Italy, and Hungary.[30]

There were also pancontinental efforts addressing prostitution around this same time and in the years preceding the antisyphilis drives. For example, the Illinois Hygiene League played an important role in the Chicago Syphilis Control Program. The league was a branch of the American Social Hygiene Association, which in turn had its roots in the American Society for Sanitary and Moral Prophylaxis—which itself emerged from the milieu that prompted the meeting of the first International Congress for Prophylaxis of Syphilis and Venereal Disease in Brussels in 1899.[31] Central to these societies was the conviction that prostitution was the root of the spread of venereal disease—ergo, the emphasis on reforming or punishing prostitutes and convincing men to practice sex only within marriage. The concerns of this international congress and subsequent national meetings and societies fostered "vice reports" in several cities, which set the tone and format for programs to address prostitution. In fact, the reports written in New York City and Chicago were central to Canada's own stance toward prostitution.[32]

Thus, the web of influences—medical, political, and moral—that enmeshed the Chicago Syphilis Control Program was woven not just within the myriad of governmental, historical, and cultural conditions of Chicago and the United States in the mid-1930s. It was part of a pattern of reasoning and structure of reform that transcended national boundaries. Moreover, it was still firmly rooted in values that were largely unchanged from the preceding century. It could be argued that the successes of the program were in part the result of the cumulative experiences of these Western countries—that Parran's five-point plan to combat syphilis was the product of almost half a century's work in devising a medical, legal, and social routine best fitted to the resources, structures, and mores of the United States of America and the countries most closely akin to it. It might also be argued that the single-mindedness of the moral analysis of syphilis, tied so closely and for so long to issues of prostitution and moral prophylaxis, made it impossible to permit a new or even coexisting ethic toward sexuality that would eschew syphilophobia, condemnation of sexual curiosity or expression outside legalized heterosexual unions, or racial bias. Finally, it could be argued still further that the ultimate failure of the Chicago—or any—Syphilis Control Program to achieve the full eradication of the disease simply reflects the anarchy of disease, that the medical obstacles alone to eradicating syphilis in 1940 would have defied any efforts at containment.

All three arguments are valid to a certain point. Certainly, the examples of syphilis control that Chicago had before it from Canada and, within its own borders, the rural South gave the program an efficient structure upon which to build. Yet by inheriting as well the biases—about sex, sex roles, and race—that underlay these programs, the Chicago Syphilis Control Program was built upon a closed set of experiences and morality that could and would not admit new approaches. At the very least, a program shaped by one moral worldview cannot address the worlds of people holding other views. At the most, an idealized, factionalized, unrepresentative—or even anachronistic—worldview will create a program that cannot hope to sustain itself in a time and among a constituency with which it is out of step.

"If You Haven't Used a Condom"

However unrealistic or inflexible the Public Health Service's position on moral prophylaxis may have been, by 1940 it was being challenged persistently, most significantly from within the program's own ranks. Whether O. C. Wenger's recommendation to begin speaking openly about the subject would have been heeded and to what effect can only be speculated upon, however, because the entry of the United States into World War II brought an abrupt end to the story of the Chicago Syphilis Control Program in its original incarnation.

War brought new priorities, if not new strategies. Cassel, in fact, ends his history of syphilis control in Canada in 1939 precisely because "the story begins to repeat itself, for the record of efforts to cope with the new crisis of VD in the armed forces is much the same as that of the First World War." And Allan Brandt observes that "anti-venereal programs developed during mobilization were closely modeled on those instituted during World War I," even that they "reflected few advances in policy from World War I."[33] It is tempting, therefore, to stop the story of the Chicago Syphilis Control Program here. A few more observations, however, must be made because, in a broader analysis of syphilis control, this return to old ways argues not just repetition but actually a setback. It is a reversion to old methods and moralities just when government, business, and medicine appeared to be on the verge of listening to new arguments and new attitudes toward syphilis control. Although the discovery of penicillin would change the nature of the problem as well as the solution for controlling venereal diseases by the end of World War II, a brief look at syphilis control in both civilian and military populations in Chicago at the start of the war demonstrates how relatively little moral and social ground had been gained.

Chicago's potential role in the war effort was considered substantial. Not only was the city one of the nation's largest manufacturers and a hub for transporting goods throughout the country, but it also was one of the largest sites for military training, with the nearby Great Lakes Training Center and several universities offering officer-training programs. In addition, it provided servicemen with "a leave and recreational center second only to New York because of its many amusement attractions."[34] City, county, and state were soon involved in several new programs directed at controlling syphilis and gonorrhea in the military, as were most communities throughout the United States. By the end of 1941, the Illinois Department of Public Health was involved in testing all Selective Service registrants. The Cook County Public Health Unit was revitalized and made an arm of the Selective Service System, setting up clinics to which infected selectees were sent for treatment "until they have had the minimum treatment required for induction into the Army."[35]

By June 1942, the Chicago Syphilis Control Program, renamed the Chicago Venereal Diseases Control Program, was pursuing its own civilian programs, but the Board of Health also entered into a new enterprise with the surgeon general and his staff. The Chicago Cooperative Project was to work with the Public Health Service to provide "assistance for and aid in the continued operation of a demonstration program designed to initiate, develop, and evaluate venereal disease control aid techniques as applicable to defense needs in war-time conditions."[36] In fact, in requesting new services and facilities for Chicago, O. C. Wenger pointedly set this relationship apart from that of Washington with the Chicago Syphilis Control Program and grounded his request for funds in a recital of the program's shortcomings. "Control measures under the old program were mostly of a conciliatory nature, whereby an infected patient was continuously persuaded to avoid exposing others and continue his treatment," he noted, an expensive, inefficient system. Now, Wenger proposed a "new and more determined program . . . based on the use of police powers of the health officer. Segregation, isolation, and quarantine of non-cooperative, infectious persons will be used when necessary."[37] Whether Wenger was truly indicting the limited achievements of the program that he had helped direct for the past four years or merely believed that the threat of a weakened national defense would (or should) permit stronger measures, he was now prepared to exercise strong legal constraints to ensure compliance with medical treatment.

Unlike the county's work with selectees, however, the Chicago Cooperative Project's contribution to the war effort was directed against prostitution. The Venereal Diseases Control Program of the Illinois Department of Health, along with similar departments in other states, signed a federal agreement in late December 1940 "to give full cooperation to a vigorous program of repression of prostitution to insure against the hazards of venereal disease among the armed forces, the war industry workers and the selective service registrants." The plan included prosecuting owners and operators of establishments where prostitution occurred, as well as taxi drivers and other people who played an "intermediary or profiting" role in prostitution. All prostitutes found to have acute gonorrhea or primary, secondary, or early latent syphilis would be hospitalized immediately, preferably voluntarily but involuntarily if necessary.[38] The federal government created the Social Protection Division of the Federal Security Agency and put at its head Eliot Ness, noted for his earlier work against vice and organized crime. In Chicago, Wenger requested and eventually received money for the purchase and operation of a hospital for the detention of "uncooperative and delinquent patients in an infectious stage who may be, or may become, a menace to others."[39] These patients were prostitutes, not servicemen, their threat to the national defense justifying the expenditure of federal money to purchase and outfit a city facility for their treatment and containment.

In Chicago, old procedures were once more brought into full play, finally with the support, at least on the federal level, that Herman Bundesen had failed to receive during the 1920s or 1930s. "It will be interesting to note what effect the newspaper stories of Doctor Bundesen's quarantining of Lane Hotel will have on future cooperation with the Board of Health by other hotel owners," Surgeon General Parran wrote in 1942 to one of his statisticians stationed in Chicago.[40] He was referring to the actions, covered in at least five local newspapers, of quarantining a house at 878 North Clark Street. The *Tribune* featured a photograph of police officer Gerald McCarthy checking the quarantine sign, and the accompanying article reported, "Chicago health officials yesterday revived the fight against venereal diseases by invoking the 32 year old law of quarantining houses of prostitution and other dwellings where infection has been discovered." Health Commissioner Bundesen spoke of past achievements but grounded his concern in the affairs of the day, saying, "Chicago has the best record in fighting venereal diseases of any city in America. . . . The department of health intends to see that this record is maintained. Special attention will be given to the protection of service men, and in this effort the commands of the armed forces are cooperating."[41] That quarantine became a regular tool in Chicago's armamentarium of vigilance is supported by a report to the surgeon general two years later: "If a hotel or beer parlor is named as a place of encounter five times in a month a registered letter goes to the manager notifying him of the situation, and if the location continues as a much-frequented place of encounter it is quarantined."[42]

The return to efforts to suppress prostitution was only one of the ways in which Chicago was asked to participate in new initiatives against syphilis and gonorrhea. New approaches to treatment appeared, based on new developments and studies in venereal disease control. Medically, the development of a five-day "intensive treatment" regime that could render noninfectious (and in some cases even cure) syphilis in less than a week made treatment less onerous and mandatory treatment more defensible. Intensive treatment at first included sulfa drugs and fever therapy for gonorrhea and, for syphilis, a five-day regime with Mapharsane (another arsenical) and bismuth. At some point, treatment switched to penicillin for syphilis and sulfa-resistant gonorrhea.[43] The Intensive Treatment Center that Wenger requested offered the new treatment, as did fifty outpatient rapid treatment centers set up around the city to treat infected persons who complied with the prescribed therapies.

While Wenger's role seems to have centered on rapid or intensive treatment, Chicago's Commissioner of Health Herman Bundesen's efforts were primarily spent creating a series of early treatment stations—"early treatment" being a euphemism for prophylaxis. Euphemism or not, by 1940 the U.S. government was finally ready to take a position that went beyond moral prophylaxis. In 1940, the Prophylactic Committee of the American Social Hygiene Association (which worked closely with the government in this war as they did in the previous one) reported to the Public Health Service that it still considered that no known chemicals would effectively prevent gonorrhea. At the same time, publications from the navy announced that condoms had been found to be more effective in preventing syphilis than were chemicals.[44] Weighing these and other reports, the Joint Committee on Prophylaxis concluded "that the condom is the most satisfactory prophylactic available at present." For their own part, religious groups reached a consensus "that health authorities have the responsibility for investigating and presenting the facts and that it is the job of the religious and related groups to present these facts in the social context which the particular group may desire."[45] At last, medical knowledge, mechanical know-how, and at least some conservative moral opinion seemed to be in accord, each identifying its own and respecting each others' realms of influence.

Bundesen proceeded with his plans. A. J. Aselmeyer, assistant chief of the Division of Venereal Diseases in the Health Service, informed Wenger that it was developing a pamphlet promoting condoms, with additional although less prominent information provided about soap and water and chemical prophylaxis.[46] Chicago soon created its own pamphlet, which it sent to the Public Health Service. In July 1942, the Chicago Venereal Diseases Control Program distributed the pamphlet "to service men in railroad stations, bus terminals, and other public places." The pamphlet, entitled "Servicemen Prevent Syphilis and Gonorrhea," advised that the "surest way" to prevent infection was to "avoid prostitutes and pickups," but in any case to use a condom during all sexual intercourse.[47] "IF YOU HAVEN'T USED A CONDOM," the pamphlet went on,

> URINATE after contact.
> USE SOAP AND HOT WATER. Wash sex organs, and nearby parts of body thoroly.
> REPORT within the hour to a Safety-first Station for early treatment.[48]

Despite the flourish and public display of these new or renewed activities, all did not proceed smoothly. "The Chicago Intensive Treatment Center appeared dark and dirty," George E. Pankhurst, an official from the Health Service, reported to Parran on July 8, 1944, "but I was told that the condition of the building is much better than when it was taken over." Nevertheless, he was uncomfortable about his visit. "Throughout the tour of the Center," he went on to say, "there was a feeling of secrecy and no one appeared to be free to explain the complete workings of the Institution, or to express his personal opinions." He sketched a scenario of repressive, quasi-legal detention of prostitutes: "On being admitted to the Center the patients' clothes are checked, and they are given uniforms. The patients have phone privileges and may call a lawyer, who often calls Doctor Bauer who refers him back to Doctor Bundesen, and so on indefinitely; and by ten days the patient is discharged from the Center before Court action has released them." The confusion of plurals and the generic *he* hides the fact that most of the patients were probably women prostitutes, whose rights now could successfully be buffaloed for the duration of treatment—treatment being so speedy that it was often completed before an appeal to the courts could even be processed. Despite the vigilance of the administrators of the hospital, however, Pankhurst concluded that the State Health Department should not continue to send patients there because people "cannot readily obtain information concerning [its] operation."

In terms of the suppression of prostitution in general, problems also arose. For example, the *Tribune* reported charges that the navy was working with some houses of prostitution in Japan, a collaboration that saddened the naval officer who worked with the government's new Social Protection Division.[49] In Illinois, it was rumored that Peoria's new mayor refused to close that city's twenty-five houses of prostitution because he and the army had "some sort of arrangement about this (keeping the houses open)."[50] Such examples reflect the similar disagreement within the Armed Services during World War I over whether suppression or regulation was the best way of dealing with prostitution, as well as the perennial struggles between federal policy and local practice on the issue.

Finally, Bundesen's early treatment centers were under fiscal pressure from the start, with the Public Health Service never funding the number of stations he wanted to create and then steadily threatening and eventually withdrawing funding because these few stations were pronounced poorly utilized.[51] Had Bundesen been given the six stations he proposed instead of only two, and funds to staff them beyond the medical students who were often the only personnel present, the stations might have achieved the visibility and accessibility to succeed. But Bundesen's problems were similar to those encountered earlier in maintaining the momentum of the dragnet stations at the height of activity and public exposure of the Chicago Syphilis Control Program. Clearly, even when officially condoned, the mechanical prevention of syphilis and gonorrhea still did not gain strong—financial or vocal—support at the federal level.

And so it went. Recast as the Chicago Venereal Diseases Control Program, Chicago Intensive Treatment Center, and Chicago Cooperative Project—transmuting undoubtedly into other programs, with budgets altering to match the tenors and economies of the times—the story could go on indefinitely and, at certain periods, as entertainingly as it did during the late 1930s. There is no arguing that the work of the Chicago Syphilis Control Program brought a record number of people under treatment, especially among the city's woefully underserved poor and predominantly African-American populations. The rates of syphilis and gonorrhea among servicemen during World War II were also dramatically lower than during World War I.[52] Steadily advancing medical and therapeutic knowledge can receive some credit, and, certainly in terms of the Chicago Syphilis Control Program, heightened public awareness must be credited for short-term gains.

Even before the entrance of the United States into world war, however, the program had difficulty maintaining public enthusiasm for its goals in the face of continued threats to funding, loss of the program's novelty, reluctance to address volatile issues such as race and sexual behavior, and the continual pressures of politics both inside and surrounding the program and its officers. Moreover, as the country felt new, greater threats to its security, it quickly returned to overtly oppressive authoritarian approaches to venereal disease control, further calling into question the purportedly enlightened approaches to sex, morality, and race that the program and its federal backers claimed. Because of the advent of antibiotics, which removed much of the fear of both death and public censure during a lengthy chemotherapy and made the need for prophylaxis theoretically less crucial, it cannot be determined whether the inroads that the Chicago Syphilis Control Program made could have sustained themselves, but circumstances surrounding the program make the possibility seem unlikely.

Old fears and prejudices die hard. They are, moreover, embedded in institutions that span decades and even national boundaries. Given such intransigence, should they be challenged or changed? Doesn't such intransigence indicate a broad consensus that ought to be honored? Chicago and the Public Health Service purportedly did not set out to change the nation's psyche or mores, but rather to improve its health. The agencies found, however, that they could not redress one without confronting the other. They chose, though, only to give lip service to that confrontation and, in actuality, to sidestep or ignore the personal, moralistic roots of the problem. More pointedly, they chose to deny that morals or individuality was present in scientific medicine. Theirs was a pragmatic approach. In retrospect, a broader pragmatics suggests that all practical solutions demand grappling head-on with the emotional, often illogical, and definitely unscientific underpinnings of a problem. Such a pragmatism demands the constant criticism and challenge of past beliefs. It searches for new ways of admitting and solving problems that are consciously and unavoidably moral.

EPILOGUE

Between Syphilis and AIDS

In 1988, Illinois became one of only two states to enact a premarital testing law for HIV status. Its enactment was followed by a 20 percent drop in applications for marriage licenses compared to 1987, as around twenty thousand couples "fled to bordering states, mainly to Indiana and Wisconsin, to get married, thus avoiding the delay and expense of an Illinois AIDS test."[1] Then, in 1989, that law was rescinded on the grounds that it was a violation of privacy and with the "realization that scarce AIDS dollars shouldn't be wasted on [low-risk] people who don't need to be tested or don't want to be tested."[2] At the same time, Illinois rescinded the Saltiel Law, offering a similar argument.

Times have changed since the days of the Chicago Syphilis Control Program. The experience of the civil rights movement and a new emphasis on individual rights have created an environment in which law and policy receive careful scrutiny to detect bias based on race, gender, sexual orientation, or other cultural or biological situations. Such sensitivity has led not only to the repeal of the premarital testing laws but also to such actions as efforts to change the Centers for Disease Control's (CDC's) definitions of AIDS in order to recognize and respond to the unique needs of women who are HIV-positive.

Syphilis, moreover, is not AIDS. Although more virulent in the past than it was in 1937, syphilis never reported AIDS's mortality rate. Furthermore, cumbersome and uncertain as it was, a cure for syphilis was available in 1937. Nevertheless, early media reporting and cultural studies of AIDS often drew comparisons with syphilis.[3] Metaphors of plague and war have studded the rhetoric of both infections.[4] Even though syphilis was curable in the 1930s, it was nevertheless repeatedly portrayed as a killer. Similarly, citizens of Chicago in 1937 usually associated syphilis with illicit sexual "promiscuity" as well as with sexual "liberation" in ways that foreshadow the associations that many people today make of HIV infection with the "gay liberation" of the 1970s.[5] The media's pronouncement of AIDS's spread from the homosexual to the heterosexual population in 1986 triggered a sudden concern among many heterosexual people who had heretofore been uninterested in the progress of the epidemic, a response that echoes Anglo-Americans' concern about syphilis spreading to them from African Americans earlier in the century.[6] Finally, Ben Reitman's plea to "make sex safe" is paralleled today in the call for "safe sex" (or "safer sex"), a slogan that shares Reitman's recognition that sexual expression is an inescapable part of human nature.

It is the human responses to syphilis and AIDS and HIV that are timeless. Fear of disease and death, discomfort with discussing sexuality, distrust of people or groups whose culture or values differ from one's own—all of these elements unite the problems posed by syphilis in 1937 and AIDS more than half a century later. These responses inform and shape the strategies, rhetoric, and policies that surround syphilis past and AIDS present.

Activism

> I don't think I'm doing anything different. I'm a writer writing. . . . I think everything I write is eminently sensible, not radical at all. . . . I guess what I'm trying to say is that I didn't start out to write anything I eventually wrote, but I'm mighty glad I was able to do so. Even if I have to be called an activist. —Larry Kramer, writer[7]

> As a physician, I have always been at least a crusader, at best a pioneer. —C. Everett Koop, U.S. surgeon general (1981–88)[8]

The passage of the Bulwinkle-LaFollette bill, which allocated $15 million over three years solely for venereal disease control, has a contemporary counterpart: the Comprehensive AIDS Resource Emergency Act of 1990, which allocated $2.9 billion over five years to the HIV epidemic.[9] The roads to those two bills, though, were quite different. The Chicago Syphilis Control Program was created primarily because a new surgeon general brought his crusade against syphilis into office with him. Thomas Parran was promoted within the Public Health Service, where his position toward syphilis control was well known and supported by President Roosevelt. C. Everett Koop, surgeon general during Ronald Reagan's presidency, on the other hand, came to public office from the private sector, a university-based pediatric surgeon and well-known spokesperson for the values of conservative Christianity. His widely publicized position against abortion prompted his selection to the post by President Reagan, who assumed—correctly—that Koop would continue to speak and advise against abortion.

At the time of his appointment, however, Koop was also ready to turn his attention to other issues: "I wanted to enter a larger arena and fight on behalf of a broader constituency," he wrote.[10] Thus, he was open to the challenges that the HIV epidemic offered, but the position he took was not the one the White House expected—or supported. Rather than condemning AIDS as proof of the immorality of homosexuality and urging a return to sex only within monogamous, heterosexual marriage, Koop spoke openly about, among other things, condoms and the hazards of anal intercourse. Would Koop have gotten federal money for research and services rolling sooner had he agreed to downplay the same elements of prevention that Parran had made taboo in the late 1930s? On the other hand, what if Koop had attacked the Reagan administration openly, left his job in defiance, and trusted to his already substantial reputation to keep him in the public eye where he could then speak more freely?

Koop writes at length about the obligation he felt to the office of surgeon general and its role to protect and educate the public.[11] Thus, in his determination to serve the *office* to which he was appointed, he may have spoken words that went against his expected "script," but he didn't act outside the limits of his prescribed role. In contrast, Ben Reitman tried to follow both the scripted words and constraints of the role assigned him by the officers of the Chicago Syphilis Control Program and the Chicago Board of Health. He finally sacrificed both script and role, however, when he felt they could not serve their purported ends. Once he did so, he was cut off from any official support for his work, and he struggled for the short remainder of his life to retrieve such backing.

Larry Kramer, the founder of New York's Gay Men's Health Crisis (GMHC) and the AIDS Coalition To Unleash Power (ACT UP), offers additional comparison. Kramer became directly involved in work around HIV when the epidemic touched his life personally—when his friends in New York's gay community became infected with the virus. Reitman's entrée to the lives of his patients with syphilis came through his professional and political interests, but although he did enjoy a closer communication with infected men and women than did Bundesen or Schmidt, Reitman was still an outsider among these people. Moreover, nothing in his personal correspondence suggests that he ever had syphilis, but even if he had, he never made a political point of his health status. Kramer, on the other hand, identified himself from the start as a gay man whose sexual history had put him at particular risk for infection. Now HIV-positive, his identification gives his words even greater urgency.

As activists, Kramer and Reitman have used some of the same strategies and rhetoric in conveying and promoting their views. In the early years of the HIV epidemic, Kramer went to Fire Island to ask for money to support services for people with AIDS/HIV and urge gay men to forswear the baths and other sex hangouts, the same direct approach that Reitman used throughout his life, whether at Bughouse Square, syphilis dragnet stations, or outside the Blackstone Theater. At a conference of gay health workers, Kramer reports that he "stood outside the conference entrance and handed out [a] letter to those attending. Then I went inside and interrupted the conference, yelling at them that they were not talking about what was *really* important." He concludes, "My reputation was now completely that of a crazy man."[12] Like Reitman's anger with what he saw as the professional complacency, self-satisfaction, and cowardice of Herman Bundesen and Louis Schmidt regarding the prevention of syphilis, Kramer railed at the GMHC when he felt it had lost its anger and focus: "I cannot for the life of me understand how the organization I helped to form has become such a bastion of conservatism and such a bureaucratic mess."[13]

Kramer also shares Reitman's distrust and despair of capitalistic institutions. Where in 1938 Reitman saw poor, predominantly African-American men and women as victims of business, medicine, and government, Kramer sees gays in that position, castigating the Church, the pope, the state, and the Supreme Court: "Because of these immoral institutions and their immoral pronouncements, we have been estranged from our families, and we have been forced to create a ghetto. We have been forced to suffer. We have been forced into AIDS."[14] Also like Reitman, Kramer's words have brought anger back on his own head. As Reitman was at times rebuked or rebuffed by prostitutes or homeless men for moralizing to them and by his medical colleagues for his radical views, Kramer was eventually rebuffed by both the GMHC and ACT UP when those groups refused to follow the directions in their work that he wished them to take. "I am still unable to resolve this fundamental problem," he said in 1987, "how to inspire you without punishing you."[15] His words reflect the combination of alienation and conviction that Reitman often expressed in his letters and reports, illustrating not only the difficulties inherent in turning personal commitment into general consensus but also the personal values that underlie most public statement.

Also like Reitman, Kramer has ultimately refused to write any script or act any role except those he has crafted for himself, a frequent source of irritation to his friends and colleagues. Kramer's refusal to compromise his views, in fact, led to his separation from both the GMHC and ACT UP. After ending his relationship with the GMHC, he wrote of the frustration that accompanied such loss: "When you are affiliated with an established organization, it's much easier to be taken seriously when talking to the media. Now I was disenfranchised, and I had lost my official soap-box."[16] An ironic reference, to one's "soap-box," in a comparison with Reitman, who felt that he could, literally, climb a soapbox anywhere. An "official" soapbox, however, whether the GMHC, the Chicago Syphilis Control Program, or the office of the surgeon general, lends weight to words. Kramer, Koop, and Reitman, although from vastly different backgrounds and political situations, repeatedly faced the same conflict between being true to themselves and being heard to good effect through the authority of an institutional voice. Their experiences—achievements and limits alike—demonstrate the complex relationships between politics and personality.

Organizational strategies of the Chicago Syphilis Control Program and groups like New York's GMHC and ACT UP also offer comparisons. From the start, the fanfare surrounding the creation and early activities of the Chicago Syphilis Control Program was a spectacle of flag-waving that at times seemed more circus than serious. Ben Reitman, for example, commented that the children carrying banners to City Hall announcing "Friday the 13th is an Unlucky Day for Syphilis" probably had little knowledge—or concern—about the purpose of their actions. On the other hand, when ACT UP/Chicago marked United Nations World AIDS Awareness Day on December 1, 1992, with a funeral procession through the downtown Loop that ended with the occupation of the governor's office in the State of Illinois Building, some participants may have been along for entertainment value but the majority knew that they were deliberately drawing attention to the plight of people with HIV infection—moreover, most either had friends who were or were themselves HIV-positive.[17]

Central to the difference between the two events is their origins. The former was orchestrated by government officials, and the latter grew from grass-roots organizations. Commentators on the epidemic have debated the significance, generally, of the gay community acting in this and other ways to "take ownership" of AIDS. Most of them say that it was a necessary step because no one else appeared inclined to help gay men, and they praise such actions for "strength[ening] . . . gay identity and gay institutions" and by creating public sympathy for gay men's courage in coming forward to fight the spread of HIV.[18] But there are also acknowledged risks in ownership. By identifying itself so closely with AIDS, the gay community has risked losing what social acceptance and political influence it has gained in the past few years.[19] Voluntarism, moreover, eventually exacts its own toll, as it is often threatened by burn-out, especially if support from other, more affluent fronts never arrives.[20] Larry Kramer does not speak alone when he fears that the success of grass-roots efforts will offer government agencies an excuse to continue their own meager level of involvement.[21]

At some level, a similar debate occurred among African-American leaders in Chicago regarding syphilis years before, when their negative experience as participants in an antituberculosis campaign led them to withdraw from the hoopla and rhetoric of the Chicago Syphilis Control Program; the costs, from their experience, were not worth the risk. Because any references to or discussions of race were forbidden by the program's officials, it consequently became impossible to create special educational programs or services tailored to the needs of the African-American (or any culturally identified) people. Today, AIDS organizations, developed first within and by the gay community, quickly expanded their reach to accommodate a more and more diverse population. Although organizations comprised of other constituencies such as Hispanics, African Americans, women, and drug users now work with people with HIV infections, many groups also work together, collaborations that have challenged many participants as people who formerly saw themselves with little in common now help each other in the most intimate aspects of living and dying.[22]

As Reitman argued that denying, downplaying, or ignoring the impact of race on the prevalence or persistence of syphilis would help neither African Americans nor the country in general, many AIDS organizations today are directly confronting differences in cultural responses to AIDS, sexuality, and drug use and designing ways to address these culturally different audiences.[23] Thus, while the New Deal may have had the fiscal and structural support to launch a campaign against syphilis that is lacking in the AIDS campaign to date, today's concepts of advocacy and empowerment motivate people in ways that were absent in the past and hold promise for sustaining the effort against the HIV epidemic—provided they serve as a model of only one of many efforts in that endeavor.

Fear

> A common response to fear is denial. I am concerned about a growing tendency of the American public to deny the AIDS problem. —C. Everett Koop[24]

> I wanted [my appeal] to be stirring and emotional and thought-provoking and scary. I didn't, and don't think that anything less than scary was, and is, occurring. —Larry Kramer[25]

That deliberated institutional programs and policies are, nevertheless, inextricable from personal beliefs, values, and emotions is demonstrated by one of the most dramatic images of the early years of AIDS: the yellow rubber kitchen gloves worn by Washington, D.C., police while working at an AIDS demonstration in 1987.[26] The inappropriate, exaggerated fear of contagion was made clear by the appearance of this innocuous household object on the hands of uniformed police. Rather than being worn to protect the police from the contagion of HIV, by which an unknown but relatively small number of protesters would have been infected, the rubber barrier (also symbolically ironic) can be read as the police officers' desire to protect themselves more generally from the "contagion" of homosexuality. The association was similar to the one Ben Reitman had seen being fostered by syphilophobia, a fear that was seldom so closely contained as to limit itself to its strict definition—fear of the disease, syphilis. Because of fears of the disease's spread, people with syphilis were often not seen or treated as separate from their disease—after all, you couldn't dismiss only a disease from a job or quarantine a micro-organism without its host.

Fear for one's personal health or safety can be played out in a number of ways, but how, publicly, to channel or predict fear's effects remains problematic. Ideally, as Surgeon General Thomas Parran and Larry Kramer would have it, fear of syphilis or AIDS would prompt two responses: recognizing one's own susceptibility to infection and then modifying one's behaviors to prevent contracting or spreading infection. The very invocation of fear, however, can have the opposite effect, as commentators in both 1938 and more than half a century later have observed. Ben Reitman argued that mass testing without job protection, the equation of extramarital sex with disease or death, or the association of syphilis with villainy or venality would drive people from seeking both testing and health care. Surgeon General Koop used the same argument five decades later when he wrote that Americans "fear this still mysterious disease. They fear its mortality rate, which is virtually 100 percent. They fear the stigma of the disease, of what other people will conclude about their behavior and their judgment. They fear the consequences of that stigma, which can be loss of a job or housing, expulsion from school, or denial of certain health or social services."[27]

Questions of how to gauge or direct fear toward positive rather than potentially self-destructive ends arose in the *Tribune*'s reporting of the gin marriage law, syphilis testing in Chicago's courts, and the statistics generated by the Central Tabulating Unit. A similar dilemma has followed the reporting of the evolution of AIDS/HIV, as the media has been condemned at various times for the absence of news coverage or for creating an "AIDS scare."[28] The scare tactics that were often part of the PHI's advertising have contemporary counterparts, for example, in the public-service poster that shows a skeleton in evening gown singing the dangers of AIDS to young women, or the cover of *Life* magazine that proclaims, "Now No One Is Safe from AIDS."[29] But matter-of-factness can put people off in its own way. For example, Reitman's reprimands from his bosses for his "vulgarity" were echoed in the Senate in 1987, when a comic strip from GMHC that demonstrated the use of a condom was condemned as "demeaning" and "repugnant."[30] The tension between presenting useful, urgently needed information and distancing people through either fear, disgust, or denial has not been resolved since the late 1930s. Fear, however, repeatedly seems to be the message of choice, an irony not lost on the media analyst Douglas Crimp, who observes, "One curious aspect of AIDS education campaigns devised by advertising agencies contracted by governments is their failure to take into account any aspect of the psych[e] but fear. An industry that has used sexual desire to sell everything from cars to detergents suddenly finds itself at a loss for how to sell a condom."[31]

Fear has been used in an attempt not only to engage people's attention but also to *sustain* their commitment through headlines, predictions, and numbers that excite dismay or anxiety. In 1937, Special Consultant Wenger blamed the drop in attendance at dragnet stations in part on the loss of the novelty of the Chicago Syphilis Control Program, and Ben Reitman worried over how to keep interest in the program alive without printing frighteningly misleading or inaccurate information in the *Tribune*. Similarly, today's observers of AIDS express concern that public interest in the needs of people with AIDS/HIV or one's personal sense of vulnerability will flag as AIDS becomes "old news."[32] They note that media coverage of the HIV epidemic has become "event driven," that is, articles generated by conferences about AIDS or planned public demonstrations—"a routine and relatively constant news item" whose predictability will avoid the debilitating "epidemic of fear" created by the media in the early years of the epidemic but may fail "to keep individuals vigilant about proper prevention."[33]

More than fifty years ago, Thomas Parran hoped that fear of contracting syphilis would lead people to adopt a "moral"—that is, married, monogamous, and heterosexual—sexuality. More recently, Larry Kramer has hoped that fear of AIDS would generate an even farther-reaching response. "If this article doesn't scare the shit out of you," he wrote in the *New York Native* in 1983, "we're in real trouble. If this article doesn't rouse you to anger, fury, rage, and action, gay men may have no future on this earth. Our continued existence depends on just how angry you can get."[34]

Although for a time Kramer urged gay men to abstain from sex until the exact cause of AIDS was known, he wrote of fear prompting other actions as well.[35] Kramer urged people to fear for and take action against the endangered lives of all gay men whose behaviors have placed them at risk for infection. He also feared for the survival of the gay community in the face of a homophobic world. "I'm afraid," he said in 1987. "I've never said that out loud to you before. I think the torrents of hate are really just beginning."[36]

Kramer's allegations of genocide are not unique; in the 1930s, Ben Reitman, Lawrence Linck, and S. W. Smith of the National Hospital Association made similar accusations about the Public Health Service's position toward African Americans and syphilis. Although such charges did not sound as loudly or openly in the 1930s as do ones from the gay community now, African-American leaders from that previous time nevertheless kept a watchful, skeptical eye on the racial representations of syphilis.

Fear for one's health becomes a double-edged sword when that fear turns from responsibility for one's own behavior to suspicion of others. Martha Gever, a media analyst, defines "fear's results [as] hatred of and discrimination against those classified as abnormal and therefore degenerate or dangerous or both."[37] The inseparability of fear from bias thus offers one explanation why fear has failed to generate an open-minded response to either AIDS or syphilis.

Moralism

> All during the AIDS epidemic I have maintained that this is one disease where the scientist and moralist walk hand in hand because they advocate the same measures, even if for different reasons. —C. Everett Koop[38]

> I've also been accused of being a "moralist"—my accusers not realizing that nothing they could say would please me more. —Larry Kramer[39]

Primary targets of moral approbation during the years of the Chicago Syphilis Control Program were extramarital, heterosexual sex and alcohol as a contributor to sexual behavior. The traditional view of female prostitutes as vectors of disease or even death has carried over to the HIV epidemic, even though sex workers in the United States do not pose a large threat in the spread of the virus.[40] Much more generally, it is public condemnation of homosexual male sex that is analogous to moral judgments about sex and syphilis in 1937. Because the first scientific articles to report the appearance of what was to be named AIDS focused on recent cases of Kaposi's sarcoma and pneumocystis pneumonia among gays in New York and Los Angeles, AIDS came to be seen as a gay disease.[41] This association was bolstered by the CDC, which labeled groups of people "at risk" for AIDS rather than "risky behaviors" that could spread the infection.[42] (In fact, sexual behaviors considered "risky" are not unique to the gay male community.)

Homosexual sex occurring in a homophobic culture keeps sexuality prominent, whether explicitly or implicitly, in the discourse around AIDS/HIV. In addition, wide social condemnation of drug use contributes even more moralistic rhetoric to today's debates than did the earlier association of alcohol with syphilis. Ben Reitman argued that pietistic injunctions to curtail or suppress certain desires or behaviors would fall on deaf ears and probably alienate the bearer of such a message. Surgeon General Koop has expressed similar awareness regarding the prevention of AIDS. In 1987, when a member of the White House staff urged Koop to talk to the public about family values and not about condoms, the surgeon general replied that "the country was not going to accept that as the president's way of addressing a national health threat."[43]

Koop's position is an interesting one. Like the earlier surgeon general, Parran, Koop held traditional moral values. "Many of my critics viewed anal intercourse as a violation of laws both spiritual and temporal," he wrote in his autobiography. "So do I." Also like Parran, Koop repeatedly insisted upon "the much better and much safer alternatives of *abstinence* and *monogamy*."[44] At the same time, though, and markedly unlike Parran, Koop also explained the particular benefits of condoms and the hazards of anal intercourse whenever he spoke about the HIV epidemic. He squared such seeming moral leniency with his conservative religion. First, he saw his role as surgeon general to be "to inform the American people about the prevention of disease and the promotion of health," a mandate that, for Koop, demanded that his personal beliefs take a back seat to public exigency.[45] Similarly, his religious beliefs argued against taking a position of moral superiority or authority. He said, "I knew in the depths of my heart that my nature, like everyone else's, was sinful, and my efforts to reform myself to no avail."[46] When Koop wrote, "I also knew that Christians are taught to separate the sin from the sinner, and treat those in need with compassion, just as Jesus did," he called for compassion, a human response as opposed to the purely scientific "objectivity" that Parran vowed was the only workable approach to syphilis.[47] In addition, Koop's belief that all people are sinners established a kind of moral equality that Parran, who held up his happy marriage and healthy children as models to the nation, did not offer.

In the final analysis, though, Koop's conclusion about AIDS echoes Parran's call for "the social regeneration of a whole people."[48] Koop aspired to—and held up as "right"—a morality that excludes all homosexual men and women or heterosexual men and women who engage in sexual activity outside marriage. He wrote, "Perhaps nothing will work except a rekindled emphasis on morality."[49] Appealing at different times to the supposedly ideologically neutral position of his office, Christian compassion, and the moral values of the "religious right," Koop's position, on the surface, may appear to be as inconsistent as Parran's avowal of scientific objectivity while creating policy that offered only moral prophylaxis to the public. Koop's acknowledgment of moral elements in his arguments, however, permitted him to respond more sympathetically and realistically to the needs of all people practicing risky behaviors. By asking himself to be morally responsible to his constituency rather than morally responsible *for* them, Koop placed himself in a position of serving rather than dictating the values of others where AIDS was concerned.

Larry Kramer also uses moral language in his analysis of AIDS, but he has tended, unlike Koop, to use it in a normative rather than conciliatory way, as he draws a direct connection between the rapid spread of HIV infection in the gay community and the deliberate celebration of sex in the late 1960s and through the 1970s. He wrote, "So rightly or wrongly—wrongly as it turned out—we decided we would make a virtue of the only thing you [i.e., the heterosexual, white, male, middle class] didn't have control over: our sexuality."[50]

Kramer's statement implies a misguided attribution of virtue and thus indirectly imposes a negative judgment on a highly visible part of gay social life during those years, a position he took even before the epidemic in his novel *Faggots*.[51] Kramer's words have angered many gays. He and other prominent spokespersons for AIDS and the gay community, such as the journalist Randy Shilts and Andrew Holleran, a novelist, have been accused of projecting their own homosexual guilt onto the discourse of the HIV epidemic.[52] AIDS, many people argue, does not—and moreover should not—mean the end to a delight in gay sexuality. For one thing, sexual behaviors can change—and have changed—without abandoning sexual activity.[53] Other writers argue the necessity of maintaining delight and self-respect in gay sexuality, saying that to do less would be to return to the days of closeted guilt or self-loathing—and isolation—that existed before the late 1960s.[54]

Homosexuality and AIDS/HIV in communities of color have added other dimensions to the moral discourse. As with African Americans and syphilis in the 1930s, African Americans and Hispanics represent a disproportionate percentage of people with HIV infection, in large part because of socioeconomic conditions.[55] Talking about homosexuality in many African-American and Hispanic communities is complicated by a homophobia that is even stronger than in many white ethnic communities.[56] In fact, in the cases of broadcast journalist Max Robinson and sports star Erwin "Magic" Johnson, images—dangerously stereotypical—of sexual prowess were deliberately promulgated by personal friends and the media to still any rumors that these men might be gay.[57]

That many men—and most women—of color have acquired the HIV virus through activities related to drug use raises another catalog of racial stereotypes, but intravenous drug-users of any race infected with HIV face additional marginalization. HIV infection went unnoticed for so long in today's drug-using population because few people cared about the health of such persons. Stereotyped distrust of drug-users' reliability has continued to make it difficult for them to receive standard treatment or to be enrolled in drug-testing programs.[58] Drug-users today share a social position with prostitutes and homeless men during the time of the Chicago

Syphilis Control Program, who were seen as, to use Ben Reitman's term, "outcasts." Fatal diseases that could be attributed to their behaviors were often interpreted as punishment befitting the crime. Providing treatment for these people—moreover, providing treatment in a way that was acceptable to them—called for a tolerance that most people at the decision-making levels in 1937 were unwilling to practice.

In general, Reitman argued repeatedly for fuller health and social services in poor communities, frank talk about sexuality and prevention of syphilis, and open discussions about the spread of syphilis in which misperceptions about race could be confronted and dispelled. Instead, officials of the program responded to racially sensitive information by suppressing all public discussion around race. The Chicago Syphilis Control Program failed in part because it did not unite Chicago's various ethnic and economic groups. Although many African Americans remained cautiously supportive of the methods of the program, other disenfranchised men and women believed that government demonstrated no immediate care for their well-being. Moreover, despite its protestations that the stigma accompanying syphilis had been removed, the *Chicago Tribune*, with its guilt-laden advertising, coverage of the passing of the Saltiel Law, and stories about Al Capone or the Reverend Philip Yarrow's son, could never completely abandon associating syphilis with "immoral" behavior. Moral judgment hung heavy in the air, despite all medical and governmental claims to the contrary.

The same divisions threaten with HIV/AIDS today, as African-American communities suspect AIDS as a deliberate white tool aimed at their genocide, their distrust in part a legacy of the Tuskegee experiments. Gay men fear the loss of their hard-won ground toward public and legal acknowledgment of their right to live as they choose. Women cite long-standing exclusion from government drug studies as a tradition that has hampered their efforts to receive the drugs and services available to many men with HIV. Intravenous drug-users infected with HIV have difficulty even within their own support systems, because there are still rehabilitation programs that are either not equipped or not willing to take on the task of helping people with HIV infection.

Yet the celebration of cultural, ethnic, or gay difference, appearing since the late 1960s, and the consequent search for an equity that accommodates these differences have made it easier for some people and groups to speak out against such firmly rooted biases. This identity also gives them strength in fighting others' fears about AIDS/HIV that are grounded in or contribute to cultural bias. Moreover, where "confidentiality" was one of the mantras of the Chicago Syphilis Control Program, many activists created by the HIV epidemic have striven to put a face—more accurately, a wealth of faces—on AIDS. Beyond the sensationalism of celebrities with AIDS, or the relative celebrity gained by people with AIDS such as Ryan White or Kimberly Bergalis, the HIV epidemic has spawned projects such as the Names Project Quilt, eloquent obituaries and eulogies, and numerous photographic and video records of identified, individual men and women whose lives have been touched by the virus.[59] Such a deliberate effort to make AIDS visible serves to spotlight human responses to AIDS other than fear or loathing, responses such as compassion, grief, love, and courage.

Paula Treichler, a linguist and cultural analyst, has called the HIV epidemic an "epidemic of signification." What we "know" about AIDS is really a nearly indistinguishable mélange of the languages and discourses of medical science, sex and sexuality, life and death, health and illness. "We cannot therefore look 'through' language to determine what AIDS 'really' is. Rather we must explore the site where such determinations *really* occur and intervene at the point where meaning is created: in language."[60] Although Treichler foregrounds language in a way that supports the premise of this entire book, I would nevertheless make one distinction between the actual etiology and pathology of disease and the pervasiveness of human discourse. The epidemic of signification is inseparable from the viral epidemic, but the "cures" for both epidemics, while closely related, must also be distinguished from each other at some points.

The physical ravages of syphilis and its physiological threat were undeniably diminished with the advent of penicillin in the 1940s.[61] At the same time, the fears and moral beliefs with which this earlier debate was fraught—its own "epidemic of signification"—did not yield as well to antibiotics. Similarly, a vaccine or cure for the HIV virus may remove the sense of immediate threat from AIDS that fuels much of the discourse around the syndrome, but the fearful, moralistic roots of that discourse, if not "cured" as well, will continue, to borrow an image that opened this book, "to smoulder in our midst." And this second, more pervasive epidemic will recur, in ever new incarnations, posing still the rhetorical, medical, and moral dilemmas that beleaguered the times of and all the participants in the Chicago Syphilis Control Program.

Notes

Introduction: Of Snakes and Snake-Killers

1. "Health Rules Advocated by Dr. Bundesen," *Chicago Tribune*, 15 August 1922, 1. Although the full name of the newspaper at that time was the *Chicago Daily Tribune*, it was generally referred to then—and is referred to now by historians—as the *Chicago Tribune*, or the *Tribune*. For the sake of such familiarity, those are the two titles that will be used here.

2. Editorial, Irving S. Cutter, "Stamp out Syphilis," *Chicago Tribune*, 1 September 1937, 12.

3. For instance, Baltimore, Maryland, had begun a vigorous treatment program following World War II, which both predated and followed some of Chicago's efforts. For a useful comparison of the two programs, see Elizabeth Fee, "Sin versus Science: Venereal Disease in Twentieth Century Baltimore," in *AIDS: The Burdens of History*, ed. Elizabeth Fee and Daniel M. Fox (Berkeley: University of California Press, 1988), 121–46.

4. Two of the best-known discussions of the cultural dimensions of metaphor are Susan Sontag, *Illness as Metaphor* (New York: Farrar, Straus and Giroux, 1977); and George Lakoff and Mark Johnson, *Metaphors We Live By* (Chicago: University of Chicago Press, 1980).

5. Information for this section has been compiled from a combination of the following sources: Allan M. Brandt, *No Magic Bullet: A Social History of Venereal Disease in the United Stated Since 1880*, expanded edition (New York: Oxford University Press, 1987); Jay Cassel, *The Secret Plague: Venereal Disease in Canada 1838–1939* (Toronto: University of Toronto Press, 1987); Edward W. Hook III and Christina M. Marra, "Acquired Syphilis in Adults," *New England Journal of Medicine*, 16 April 1992, 1060–69; Catherine M. Hutchinson and Edward W. Hook III, "Syphilis in Adults," *Medical Clinics of North America* 74 (November 1990): 1389–416; Franklyn N. Judson, "Gonorrhea," *Medical Clinics of North America* 74 (November

1990): 1353–66; and Robert L. Murphy, "Sexually Transmitted Diseases," in *The Biologic and Clinical Basis of Infectious Diseases*, 4th ed., ed. Stanford T. Shulman, John P. Phair, and Herbert M. Sommers (Philadelphia: W. B. Saunders Company, 1992), 238–68. Only references to material quoted from these sources or references to additional sources will be cited in this section.

6. Judson, "Gonorrhea," 1353.

7. Personal communication from Kahn to Ben Reitman, Venereal Disease Control [VDC] Report 152, n.d., probably between 17 and 20 November 1937, in the papers of Benjamin Lewis Reitman, M.D., Special Collections Department, University of Illinois at Chicago [hereafter referred to as BR], Supplement II, folder 30.

8. G. G. Taylor, "The Treatment of Syphilis," *Chicago V.D. Bulletin* 1 (July 1937): 7–8. A copy of this report in contained in BR, Supplement II, folder 19.

9. VDC Report 225, 22 February 1938, BR, Supplement II, folder 40.

10. VDC Report 141, n.d., probably between 1 and 5 November 1937, BR, Supplement II, folder 29.

11. The history of sulfanilamide and the debate surrounding it is extensive. For representative discussion of this debate in Reitman's papers, see VDC Report 92, n.d., between 17 and 22 July 1937, in a private collection of Reitman's papers owned by Ruth and Joel Surgal, Chicago, Illinois [hereafter referred to as SUR]. In BR, see VDC Reports 96, 27 July 1937, Supplement II, folder 23; 108, 2 September 1937, Supplement II, folder 24; 145, 9 November 1937, Supplement II, folder 29; 273, 23 April 1938, Supplement II, folder 46; 13, 16 August 1938, Supplement II, folder 54; and 43, 26 September 1938, Supplement II, folder 58.

12. The term *sexually transmitted disease* attempts to foil this connotation, although whether any form of the word *sex* can be perceived neutrally could probably be questioned. For purposes of historical consistency rather than any moral judgment, I shall use the older term here in speaking of syphilis and gonorrhea before the 1970s.

13. United States Public Health Service, *Annual Report of the Surgeon General of the Public Health Service of the United States for the Fiscal Year 1936* (Washington, D.C.: United States Government Printing Office, 1936), 11.

14. United States Public Health Service, *Annual Report of the Surgeon General of the Public Health Service of the United States for the Fiscal Year 1937* (Washington, D.C.: United States Government Printing Office, 1937), 115.

15. Herman R. Bundesen to Thomas Parran, 25 May 1936, National Archives [hereafter referred to as NA], Record Group 90, Box 27, folder 0135.

16. See also letters from Thomas Parran to Herman Bundesen, 2 June 1936, and Frank Jirka to Herman Bundesen, 13 May 1936, both in NA, Record Group 90, Box 27, folder 0135.

17. James H. Jones, *Bad Blood: The Tuskegee Syphilis Experiment* (New York: Free Press, 1981), 55–58, 137.

18. R. A. Vonderlehr to Thomas Parran, 11 December 1936, NA, Record Group 90, Box 27, folder 0425.

19. For more on the lively career of Louis Schmidt, see William K. Beatty, "Louis E. Schmidt: Pioneer and Outcast," in *Proceedings of the Institute of Medicine of Chicago* (in press).

20. *Annual Report of the Surgeon General 1937*, 113.

21. O. C. Wenger, *An Evaluation of the Chicago Syphilis Control Program after One Year* (Chicago, 15 April 1938), 2, NA, Record Group 90, Box 29, no folder. Although subsequent reports about the Chicago Syphilis Control Program were issued as annual reports from the program itself, this first report bears Wenger's name as author.

22. Transcript of "Meeting Called by the Honorable Edward J. Kelly, Mayor of the City of Chicago, in connection with Chicago's Venereal Disease Control Program," NA, Record Group 90, Box 29, folder 0425.

23. Wenger, *An Evaluation*, 19.

24. Suzanne Poirier, "The Patients and Poems of Dr. Ben Reitman—Clap Doctor of Chicago," *Proceedings of the Institute of Medicine of Chicago* 39 (Fall 1986): 124–30f.; and "Emma Goldman, Ben Reitman, and Reitman's Wives: A Study in Relationships," *Women's Studies* 14 (1987): 277–97.

Chapter 1: Taking the Pledge

1. "Meeting Called by the Honorable Edward J. Kelly." The phrase "conspiracy of silence" was used by Parran in much of his writing about syphilis at that time.

2. Stanley Armstrong, "Hail New Law as Powerful Aid to Syphilis War," *Chicago Tribune*, 25 June 1937, 8.

3. Included in Wenger, *An Evaluation*.

4. Ibid.

5. "Meeting Called by the Honorable Edward J. Kelly."

6. Brandt, *No Magic Bullet*, 8.

7. Ibid., 23–24.

8. Ibid., 20–27.

9. For a discussion of the rising status and weight of the medical profession, see Paul Starr, *The Social Transformation of American Medicine* (New York: Basic Books, 1982).

10. Brandt *No Magic Bullet*, 31–38.

11. Ibid., 52–55.

12. Ibid., 60–62.

13. Ibid., 96–116; quotation from 114.

14. Ibid., 122–24.

15. Ibid., 125–29.

16. Thomas Parran, "The Next Great Plague to Go," *Survey Graphic* 25 (July 1936): 405–11; "Why Don't We Stamp Out Syphilis," *Reader's Digest* (July 1936): 65–73; and *Shadow on the Land: Syphilis* (New York: Reynal and Hitchcock, 1937), see esp. 247.

17. Wenger, *An Evaluation*, vii.

18. *Annual Report of the Surgeon General 1936*, 11.

19. Parran, *Shadow on the Land*, 31.

20. Elizabeth Fee, "Sin versus Science" and "Urges Inclusion of Gonorrhea in Syphilis Drive," *Chicago Tribune*, 12 June 1937, 12.

21. "Urges Inclusion of Gonorrhea in Syphilis Drive."

22. William O'Neil, "Asks for Added Health Powers in Syphilis War," *Chicago Tribune*, 9 June 1937, 13.

23. Stanley Armstrong, "All Chicagoans Urged to Take Syphilis Tests, *Chicago Tribune*, 14 June 1937, 1.

24. "Presbyterian Church Joins War on Syphilis," *Chicago Tribune*, 20 June 1937, sec. 3–1.

25. "Health Service Hails Chicago's Syphilis Battle," *Chicago Tribune*, 26 July 1937, 4.

26. Arthur Evans, "Launch City-Wide Poll on Syphilis Tests," *Chicago Tribune*, 24 July 1937, 1, 10.

27. "Lions Members Urged to Aid War on Syphilis," *Chicago Tribune*, 24 July 1937, 10.

28. "Food Handlers Join Campaign to End Syphilis," *Chicago Tribune*, 16 August 1937, 16. Similar articles in the *Tribune* on this subject include "Women Mobilize Cook County Clubs for Syphilis Tests," 19 October 1938, 3; "Club Members Will Take Free Syphilis Tests," 1 December 1937, sec. 1–10; and "Cook County Council of Legion Backs War on Syphilis," 5 August 1937, 5.

29. "Syphilis Tests Taken by 65 Percent of Tribune Staff," *Chicago Tribune*, 2 April 1938, 6.

30. Arthur Evans, "Medical Society Backs Drive for Syphilis Tests," *Chicago Tribune*, 31 July 1937, 3, and "Dental Society Enlists 3,000 in War on Syphilis," *Chicago Tribune*, 3 August 1937, 9.

31. Arthur Evans, "Younger Women Willing to Take Syphilis Tests; Poll Results Encouraging, Says Dr. Bundesen," *Chicago Tribune*, 27 August 1937, 16, and "Council Votes $50,000 in War to End Syphilis," *Chicago Tribune*, 6 August 1937, 13.

32. "City-Wide Tests for Syphilis to Begin Tomorrow," *Chicago Tribune*, 31 August 1937, 11.

33. "200,000 Signify Desire to Take Syphilis Tests," *Chicago Tribune*, 21 August 1937, 3.

34. R. A. Vonderlehr to W. B. Greene, 10 December 1983, NA, Record Group 90, Box 27, folder 0425.

35. Wenger, *An Evaluation*, 20.

36. Thomas Neville Bonner, *Medicine in Chicago, 1850–1950: A Chapter in the Social and Scientific Development of a City* (Madison: American History Research Center, 1957). The following discussion is drawn from the chapter "Public Health Work in Chicago," 175–98, unless otherwise noted.

37. Stanley Armstrong, "Hail New Law as Powerful Aid to Syphilis War," *Chicago Tribune*, 25 June 1937, 8.

38. Conrad Seipp, "Organized Medicine and the Public Health Institute of Chicago," *Bulletin of the History of Medicine* 62 (1988): 429–49. The following discussion of the Public Health Institute will be based on this article unless noted otherwise.

39. Special Report 18, 3 March 1939, BR, Supplement II, folder 21.

40. Patricia Spain Ward, "Rachelle Yarros," in *Notable American Women, 1607–1950*, ed. Edward T. James et al. (Cambridge: Belknap Press of Harvard University Press, 1971): 3:693–94.

41. Chicago Board of Health, *Report of the Division of Social Hygiene, July 1, 1925 through June 30, 1926* (Chicago), 4. A copy of this report is contained in BR, Supplement II, folder 19.

42. Ibid., 4.

43. "Health Rules Advocated by Dr. Bundesen," *Chicago Tribune*, 15 August 1922, 1.

44. Chicago Board of Health, *Division of Social Hygiene 1926*, 27.

45. Bonner, *Medicine in Chicago*, 193–94.

46. Ibid., 194.

47. A. A. Dornfeld, *Behind the Front Page: The Story of the City News Bureau of Chicago* (Chicago: Academy Chicago Publishers, 1983), 139.

48. "Children Parade Loop in Crusade against Syphilis," *Chicago Tribune*, 14 August 1937, 4 and back page.

49. "Airplane Calls City Attention to Syphilis Fight," *Chicago Tribune*, 15 August 1937, sec. 1–17.

Chapter 2: The Clap Doctor of Chicago

1. VDC Report 101, 14 August 1937, BR, Supplement II, folder 23. In the original quotation, Reitman mistakenly referred to Carsten as Carstens. In this and the following quotations, such errors, misspellings, and obvious typographical errors have been corrected.

2. Elmer Gertz and Bruce Milton, *Clown of Glory*, in BR, Supplement III, folder 6. This quotation from an unpublished biography of Reitman was used with the permission of Elmer Gertz.

3. *Following the Monkey*, 4, BR, folder 11. Internal references in this manuscript indicate that Reitman was writing the major part of his autobiography in 1925, but the folder that holds this document is labeled 1933. His daughter, Mecca Carpenter,

notes in personal correspondence that he was still working with these papers in the late 1930s, so some of his representations of his thoughts and actions in earlier years may be colored by the impression he wished to leave of his life later on. All of the information in this chapter is drawn from this autobiography unless otherwise cited. Only direct quotations will be noted.

4. The story of Reitman's hobo years is best told in a biography of Reitman by Roger Bruns, *The Damndest Radical: The Life and World of Ben Reitman, Chicago's Celebrated Social Reformer, Hobo King, and Whorehouse Physician* (Urbana: University of Illinois Press, 1987).

5. *Following the Monkey*, 111, BR, folder 11.

6. VDC Report 325, 1 July 1938, BR, Supplement II, folder 52.

7. *Following the Monkey*, 144, BR, folder 11.

8. Patricia Spain Ward discovered papers to this effect while researching the history of the University of Illinois at Chicago's College of Medicine. She reports that a number of students were dismissed each year from this (and other universities) on similar charges; apparently it was not uncommon for students to hire someone to sit for one or more of their examinations.

9. *Following the Monkey*, 163, BR, folder 11.

10. Ibid.

11. Ibid., 176.

12. Bruns, *Damndest Radical*, 21–23.

13. *Following the Monkey*, 189, BR, folder 11.

14. Bruns, *Damndest Radical*, 54–55.

15. The summary of Goldman's life presented here draws generally from two major biographies of Emma Goldman and Goldman's own autobiography. The biographies are Candace Falk, *Love, Anarchy, and Emma Goldman: A Biography* (New York: Holt Rinehart and Winston, 1984); and Alice Wexler, *Emma Goldman: An Intimate Life* (New York: Pantheon, 1984). Goldman's biography, published in two volumes, is *Living My Life* (1931; rpt. New York: Dover, 1970). Only quotations or specific references will be cited.

16. Goldman, *Living My Life*, 1:421.

17. Ben Reitman to Emma Goldman, 1914, BR, Supplement II, folder 96.

18. Wexler, *Emma Goldman*, 85.

19. Goldman, *Living My Life*, 2:514.

20. Ben Reitman to Hutchins and Neith Hapgood, 27 February 1940, BR, Supplement II, folder 149.

21. Goldman, *Living My Life*, 2:527.

22. Emma Goldman to Ben Reitman, 9 March 1926, BR, folder [1–2] 83–26.

23. Martindale is identified as a suffragist in an unidentified newspaper clipping entitled "Dr. Reitman Finds Long Lost Daughter Here." The article refers to Reitman's daughter, Helen, who hopped a train in her father's best style and visited him around 1924. The article is found in BR, Supplement II, folder 62.

24. "Reitman Jars Farewell Party to Anarchists, Renounces Emma Goldman and Talks to Golden Rule," unidentified newspaper article in BR, Supplement II, folder 62.

25. *Following the Monkey*, 369–370, BR, folder 13.

26. Ibid., 370–71, folder 13.

27. 7 February 1934, BR, folder 8. Two versions of this poem exist, the one quoted here is typed on Pfister's letterhead. Pfister had offices in the same building as Reitman and also saw Reitman socially (Brutus Reitman to Ben Reitman, 26 May 1935, SUR). In SUR, a handwritten copy of the same poem, presumably also by Pfister but not signed, is written as a letter to Reitman from St. Louis. It is also dated 7 February 1934, but the words are not exactly the same. Instead of the last two lines of the typescript poem, the handwritten one reads: "All very positive, all so sure, / All trying to justify themselves and their acts. / And in their great desire to explain and justify, / They talk on and on / Until the Master gives the cue for the next to begin. / I say Master rather than Judge—for he does not judge."

28. *Following the Monkey*, 371, BR, folder 13.

29. Ibid., 373, folder 13.

30. Personal communication with Patricia Spain Ward, based on her research on the history of the College of Medicine at the University of Illinois.

31. *Following the Monkey*, 395–96, folder 13.

32. Brandt, *No Magic Bullet*, 40–41.

33. *Following the Monkey*, 374, BR, folder 13.

34. Ibid.

35. Ibid., 404, folder 13.

36. Document signed by L. E. Schmidt and W. A. Evans, SUR.

37. *Following the Monkey*, 415–16, folder 13.

38. Ibid.

39. Ibid.

40. Bruns, *Damndest Radical*, 273.

41. *Following the Monkey*, 412, BR, folder 13.

42. Margaret Parker to Ben Reitman, the first with no date, followed by one dated 17 November 1924, BR, folder 8.

43. Leona Clark to Ben Reitman, printed in VDC Report 146, 10 November 1937, BR, Supplement II, folder 29.

44. VDC Special Report, n.d., between Reports 11 and 12, before 18 February 1937, SUR.

45. See also VDC Report 68, 21 May 1937, and VDC Report 76, n.d., between 21 May and 19 June 1937, both in SUR. In BR, see VDC Report, 173, n.d. but probably 14 December 1937, Supplement II, folder 32.

46. *Following the Monkey*, 413–14, BR, folder 13.

47. Parran, *Shadow on the Land*, 221.

48. References to a study of acid-based chemicals are made in the *Annual Report*

of the Surgeon General 1936 (120), *Annual Report of the Surgeon General 1937* (118), and United States Public Health Service, *Annual Report of the Surgeon General of the Public Health Service of the United States, for the Fiscal Year 1938* (Washington, D.C.: United States Government Printing Office, 1938), 135.

49. R. A. Vonderlehr to O. C. Wenger, 27 February 1937, NA, Record Group 90, Box 29, folder 0425.

50. R. A. Vonderlehr to O. C. Wenger, 12 May 1938, and Thomas Parran to O. C. Wenger, 6 May 1938, both in NA, Record Group 90, Box 29, folder 0425; see also chapter 6.

51. Duties of Investigator Ben L. Reitman, M.D., SUR. Although this document is unsigned and does not appear with a letterhead, Reitman's references in his Venereal Disease Control Reports indicate that, whoever created this itemized list, it represents work that Reitman had been officially appointed to do.

52. VDC Report 154, 20 November 1937, BR, Supplement II, folder 30. This "one man show," however, was also sustained by the hard work of Reitman's secretary and sometimes-mistress, Eileen O'Connor. Reitman wrote about her, "[She] has had fifteen years experience in hospital management, the care of the sick and the prevention of disease and the education of doctors and nurses. She attends to a very large correspondence, clips the newspapers and magazines for articles on Syphilis, and acts as a consultant for the hoboes, students, newspaper men, sexologists, whores, pimps and homos who come up for a chat or consultation. And writes all the letters and reports" (VDC Report 20, 3 March 1937, SUR). Reitman probably means *types* for *writes*, although she certainly could have fleshed out many of these reports from his directions. The rhetorical style of the reports, however, is too much Reitman's for her to have written them alone.

53. VDC Report 154.

54. Ben Reitman to Thomas Parran, VDC 69, 23 May 1937, SUR.

55. Ben Reitman to Reuben Kahn, 22 December 1937, SUR.

56. Ben Reitman to "Marie Belle, and Jan," 3 April 1937, SUR.

57. Duties, SUR.

58. VDC Report 124, n.d., probably early October 1937, BR, Supplement II, folder 26. (There are two reports in this folder marked 124; this is the second of them.)

59. Ben Reitman to W. A. Evans, 8 April 1937, contained in the personal collection of Mecca Carpenter; VDC Report 20, 3 March 1937.

60. Ben Reitman to "My Dae," 19 March 1938, BR, Supplement II, folder 71.

61. Ben Reitman to Lawrence Linck, 24 February 1938, SUR.

62. Executive Secretary of the League to Ben Reitman, 5 May 1938, BR, Supplement II, folder 71.

63. Samuel Moss, personal communication, November 4, 1983.

64. Gertz and Milton, *Clown of Glory*, BR, Supplement III, folder 8.

Chapter 3: Taking the Plunge

1. "Marriage Test Bill Signed," *Chicago Tribune*, 24 June 1937, 6.
2. Brandt, *No Magic Bullet*, 14, 147.
3. "Marriage Test Bill Signed."
4. Percy Ward, "Illinois Takes Lead in Law for Marriages," *Chicago Tribune*, 10 July 1937, 3.
5. "Marriage Test Bill Signed."
6. "Hail New Law as Powerful New Aid to Syphilis War," *Chicago Tribune*, 25 June 1937, 8.
7. "Seeks Opinion on Sections of Marriage Law," *Chicago Tribune*, 25 June 1937, 9.
8. "Rush Marriage Bureau to Beat Hygienic Law," *Chicago Tribune*, 27 June 1937, sec. 1–13.
9. "1,119 Couples Get Licenses in Day to Beat Marriage Law," *Chicago Tribune*, 29 June 1937, 5.
10. Virginia Gardner, "Jam Marriage Bureau to Beat Test Deadline," *Chicago Tribune*, 1 July 1937, 1.
11. "Rush Marriage Bureau to Beat Hygienic Law."
12. Gardner, "Jam Marriage Bureau."
13. Unexpectedness, however, could also be questioned; Brandt reports that states which preceded Illinois in passing hygienic laws experienced similar drops in numbers of licenses issued after passage of the new laws (*No Magic Bullet*, 149).
14. "Marriage Law Is News," *Chicago Tribune*, 4 July 1937, sec. A-7.
15. "Teach Marriage Law's Value, Say Foes of Syphilis," *Chicago Tribune*, 25 November 1937, sec. 3–2.
16. "Illinois Takes Lead in Law for Marriages; New Act Puts Curb on Gin Marriages," *Chicago Tribune*, 10 July 1937, 3; "Gin Marriage Law Blocks Forty Couples at License Bureau," *Chicago Tribune*, 11 July 1937, sec. A-7.
17. Marcia Winn, "Marriage Law Fugitives Flock to Crown Point," *Chicago Tribune*, 12 July 1937, 11.
18. "Dodge Marriage Law and Laugh in Crown Point," *Chicago Tribune*, 1 August 1937, sec. 1–5.
19. "Teach Marriage Law's Value."
20. Doris Lockerman, "To Wed or Not? A Coin Decides at Crown Point," *Chicago Tribune*, 2 August 1937, 3, emphasis added.
21. "Wedding Clerks Paper Doves Net $60,000 a Year," *Chicago Tribune*, 8 August 1937, sec. A-3.
22. "Dust off 1852 Law to Block Gin Marriages," *Chicago Tribune*, 14 October 1937, 19.
23. "Senate Passes Marriage Test for Wisconsin," *Chicago Tribune*, 19 June 1937, 7; "Michigan Puts New Marriage Law in Effect," *Chicago Tribune*, 23 October 1937,

12; "Illinois Couple Denied Marriage Permit; Drunk," *Chicago Tribune*, 2 October 1937, 18.

24. "Link Three Broken Marriages to Indiana Justice," *Chicago Tribune*, 14 August 1937, 1; "Two More Weddings at Crown Point Wind up in Court," *Chicago Tribune*, 18 August 1937, 13; "Fifth Crown Point Wedding Strikes Snag in a Week," *Chicago Tribune*, 20 August 1937, 9; "Four More Indiana Marriages End in Chicago Court," *Chicago Tribune*, 26 August 1937, 5.

25. "Couple Dodge Marriage Law; No Annulment!" *Chicago Tribune*, 1 October 1937, 1.

26. Lloyd Wendt, *Chicago Tribune: The Rise of a Great American Newspaper* (Chicago: Rand McNally, 1979), 607, 725.

27. Stanley Armstrong, "*Tribune* Pioneer in Open Fight on Syphilis Plague," *Chicago Tribune*, 18 June 1937, 19.

28. Wendt, *Chicago Tribune*, 553, 607; Arthur Evans, "Launch City-Wide Poll on Syphilis Tests," *Chicago Tribune*, 24 July 1937, 1.

29. "Publicity for the Syphilis Campaign," editorial, *Chicago Tribune*, 26 January 1938, 10.

30. "Marriage Test Bill Signed; In Force on July 1," *Chicago Tribune*, 24 June 1937, 6; "Hodes Proposes Syphilis War on National Front," *Chicago Tribune*, 19 July 1938, 10.

31. "Meeting Called by the Honorable Edward J. Kelly."

32. Wenger, *An Evaluation*, 10.

33. Quoted in John J. McPhaul, *Deadlines and Monkeyshines: The Fabled World of Chicago Journalism* (Westport: Greenwood Press, 1962), 291.

34. Reitman's reports note much of the news coverage of the program's activities. He observes articles questioning the social or economic goals of the program in both papers, as well as stories acknowledging and educating readers about chemical prophylaxis at a time when the *Tribune* was picking its way carefully around the issue. Reitman's references to coverage in other Chicago newspapers generally express much the same frustration that he voiced about the *Tribune*. His observation of *Tribune* coverage, however, occurs easily six to seven times more often than of other papers, indicating (because Reitman quite thoroughly commented on coverage of the program in local media) the greater extent to which the *Tribune* covered the program's work.

35. The histories and memoirs of Chicago newspapers are many and lively. The most detailed account of the *Tribune* is found in Wendt's *Chicago Tribune*. Wendt's enthusiasm can be balanced by (and, in turn, he can balance) other authors of similar books. These other accounts include books by Dornfeld (*Beyond the Front Page*) and McPhaul (*Deadlines and Monkeyshines*), as well as William T. Moore, *Dateline Chicago: A Veteran Newsman Recalls Its Heyday* (New York: Taplinger Publishing Company, 1973).

36. Wendt, *Chicago Tribune*, 32, 357.

37. Ibid., 356–58.

38. Armstrong, "Hail New Law."

39. Nearly the complete series of articles that ran in the *Tribune* has been collected and reprinted by the Bureau of Investigation of the American Medical Association as *"Men's Specialists": Some Quacks and Their Methods* (Chicago, Illinois: American Medical Association, n.d.).

40. *"Men's Specialists,"* 3–5.

41. Ibid., 9.

42. Dornfeld, *Beyond the Front Page*, 3.

43. Wendt, *Chicago Tribune*, 358–59.

44. Ibid., 391–92.

45. Ibid., 386.

46. The story of the Everleigh House and its proprietresses, sisters Minna and Ada Everleigh, has been told by Charles Washburn in *Come into My Parlor: A Biography of the Aristocratic Everleigh Sisters of Chicago* (New York: Knickerbocker Press, 1934); and in a fictionalized account by Wallace Irving, *Celestial Bed* (New York: Dealcorte Press, 1987).

47. Wendt, *Chicago Tribune*, 401–3.

48. VDC Report 83, n.d., between 27 June and 3 July 1937, SUR.

49. VDC Report 131, n.d., but probably a day or so before 20 October 1937, BR, Supplement II, folder 28.

50. VDC Report 197, 19 January 1938, BR, Supplement II, folder 35.

51. VDC Report 127, 10 October 1937, BR, Supplement II, folder 27.

52. VDC Report 149, n.d., probably a day or two before 17 November 1937, BR, Supplement II, folder 30.

53. VDC Report 164, 3 December 1937, BR, Supplement II, folder 31.

54. Parran, *Shadow on the Land*, 57.

55. VDC Report 304, 1 June 1938, BR, Supplement II, folder 50.

56. VDC Report 325, 1 July 1938, BR, Supplement II, folder 52.

57. VDC Report 25, 2 September 1938, BR, Supplement II, folder 56. Reitman began renumbering his reports from number 1 shortly after the start of reporting year 1938–39.

58. "Judge Shifts Marriage Mill into Reverse," *Chicago Tribune*, 25 March 1938, 3.

59. "Injunction Puts Stop to Indiana Marriage Mill," *Chicago Tribune*, 8 November 1937, 9; "Indiana Pastors Bar Patrons of Marriage Mills," *Chicago Tribune*, 24 October 1937, sec. 1-13.

60. "Kill Indiana Marriage Mills," *Chicago Tribune*, 12 January 1938, 1; "Exit The Marriage Mill," editorial, *Chicago Tribune*, 13 January 1938, 8; "Marriage Tests Bring Quick Drop in Divorce Rates," *Chicago Tribune*, 13 January 1938, 9.

61. Brandt, *No Magic Bullet*, 147–48.

62. "Two States Guard against Evasion of Blood Tests," *Chicago Tribune*, 3 October 1938, 6.

63. "Sale of Wedding Licenses Still in Slump in County," *Chicago Tribune*, 1 April 1938, 13.

64. Wendt, *Chicago Tribune*, 373; see also McPhaul, *Deadlines and Monkeyshines*, 9.

65. McPhaul, *Deadlines and Monkeyshines*, 9–10.

66. Lawrence J. Linck, untitled, undated speech, 2–3, copied in BR, Supplement II, folder 22.

67. McPhaul, *Deadlines and Monkeyshines*, 51n.

68. Albert E. Russell to Thomas Parran, 2 April 1938, NA, Record Group 90, Box 29, folder 0425.

Chapter 4: Taking the Rap

1. Arthur Evans, "Younger Women Willing to Take Syphilis Tests," *Chicago Tribune*, 27 August 1937, 16.

2. Stanley Armstrong, "Hail New Law as Powerful Aid to Syphilis War," *Chicago Tribune*, 25 June 1937, 8.

3. VDC Report 68, May 1937, SUR.

4. "Judges Demand Legal Curb for Syphilis Carrier," *Chicago Tribune*, 27 July 1937, 4.

5. Editorial, "Carriers of Venereal Disease," *Chicago Tribune*, 29 July 1937, 10.

6. Arthur Evans, "Statutes Allow Forced Care of Syphilis Carrier," *Chicago Tribune*, 11 August 1937, 4.

7. Arthur Evans, "Judge Orders Syphilis Tests for Prisoners; Plan to Compel Cure of Arrested Men," *Chicago Tribune*, 17 August 1937, 12.

8. Arthur Evans, "First Men Take Syphilis Tests at Court Order; Judge Starts System on Ten Prisoners," *Chicago Tribune*, 18 August 1937, 13.

9. Arthur Evans, "Forced Syphilis Test Revelation Spurs Campaign," *Chicago Tribune*, 24 August 1937, 7.

10. "Adopt Kahn Test for Use in City's War on Syphilis," *Chicago Tribune*, 3 September 1937, 13.

11. Arthur Evans, "Find Fifty Victims of Syphilis in One Court Test; 150 Sent to Clinic; a Third Infected," *Chicago Tribune*, 19 August 1937, 1.

12. The Vice Commission of Chicago, *The Social Evil in Chicago: A Study of Existing Conditions with Recommendations by the Vice Commission of Chicago* (Chicago, 1911), 27.

13. Vice Commission of Chicago, *The Social Evil*, 25.

14. Ibid., 27–28.

15. Chicago Board of Health, *Division of Social Hygiene 1926*, 27.

16. Ibid., 24.

17. Ibid., 25–26, emphasis added.

18. Ben L. Reitman, *Sister of the Road: The Autobiography of Boxcar Bertha as told to Dr. Ben Reitman* (New York: Sheridan House, 1937).

19. Reitman, *Sister of the Road*, 187–88.

20. Ibid., 191.

21. "Charlotte," *Outcast Narratives*, number 87, BR, Supplement II, folder 9.

22. "Evelyn," *Outcast Narratives*, number 77, BR, Supplement II, folder 9.

23. VDC Report, 95, n.d., probably written a day or two before 27 July 1937, BR, Supplement II, folder 23.

24. VDC Report 82, 27 June 1937, SUR.

25. VDC Report 108, 2 September 1937, BR, Supplement II, folder 24.

26. VDC Report 62, n.d., between 3 March and 20 May 1937, SUR.

27. VDC Report 108.

28. "Put Gonorrhea Quarantine on West Side Flat," *Chicago Tribune*, 16 October 1937, 2f.

29. Wenger, *An Evaluation*, 9.

30. Chicago Syphilis Control Program, *Annual Report of the Chicago Syphilis Control Program—July 1, 1939–June 30, 1940*, 53. A copy of this report is contained in BR, Supplement II, folder 18.

31. VDC Report 197, 19 January 1938, BR, Supplement II, folder 36.

32. VDC Report 56, n.d., between 3 March and 20 May 1937, SUR.

33. Ibid.; VDC 86, 5 July 1937, SUR.

34. Parran, *Shadow on the Land*, 224–25.

35. Ibid., 267, emphasis added.

36. Ibid., 221.

37. Ibid., 222–23.

38. Stanley Armstrong, "All Chicagoans Urged to Take Tests," *Chicago Tribune*, 14 June 1937, 1.

39. "Ask 271 Million for U.S. Fight on Syphilis Plague," *Chicago Tribune*, 15 February 1938, 11.

40. Stanley Armstrong, *Chicago Tribune*, 18 June 1937, 19; *Chicago Tribune*, 18 June 1937, 19.

41. Arthur Evans, "Council Votes $50,000 in War to End Syphilis," *Chicago Tribune*, 6 August 1937, 13.

42. "Young Yarrow and Woman Face Physical Tests," *Chicago Tribune*, 4 September 1937, 16.

43. Evans, "First Men Take Syphilis Tests."

44. VDC Report 104, 18 August 1937, BR, Supplement II, folder 23.

45. VDC Report 105, 21 August 1937, BR, Supplement II, folder 23.

46. VDC Report 104, emphasis added.

Chapter 5: Dragnets

1. Arthur Evans, "Double Chicago Syphilis Tests in Three Weeks," *Chicago Tribune,* 20 August 1937, 9; and "Forced Syphilis Test Revelation Spurs Campaign," *Chicago Tribune,* 24 August 1937, 7.

2. Arthur Evans, "Report on Tests Spurs Campaign to End Syphilis," *Chicago Tribune,* 26 August 1937, 10.

3. VDC Report 60, n.d., between 3 March and 17 May 1937 (its number suggests an early May date), SUR.

4. "City-Wide Tests For Syphilis to Begin Tomorrow," *Chicago Tribune,* 31 August 1937, 11.

5. "Tests to Start Today in Fight to End Syphilis," *Chicago Tribune,* 1 September 1937, 4; "Adopt Kahn Test for Use in City's War on Syphilis," *Chicago Tribune,* 3 September 1937, 13.

6. "Response Heavy as Tests Start in Syphilis War," *Chicago Tribune,* 2 September 1937, 4.

7. Louis E. Schmidt, "Qualifications of a Venereal Disease Clinician," *Chicago V.D. Bulletin* 1 (July 1937): 4. A copy of this report is contained in BR, Supplement II, folder 19.

8. Wenger, *An Evaluation,* 10 and exhibit III.

9. Ibid., 33.

10. Ibid., 12.

11. Ibid., 12–13.

12. Ibid., 13.

13. Ibid., 13–14.

14. Ibid., 14.

15. "War on Syphilis Spurs Reporting of Cases in City," *Chicago Tribune,* 9 November 1937, 5.

16. "19,000 Treated in City, Syphilis Survey Shows," *Chicago Tribune,* 17 November 1937, 16.

17. Wenger, *An Evaluation,* 14 and exhibit VI.d; David C. Elliott to Thomas Parran, "Narrative Report of Activities, July 6–August 13, 1937," NA, Record Group 90, Box 28, folder 1850.

18. Wenger, *An Evaluation,* 14–15.

19. "War on Syphilis Gains Headway, Statistics Show," *Chicago Tribune,* 10 December 1937, 21.

20. Wenger, *An Evaluation,* 15.

21. Ibid.

22. VDC Report 9, n.d., but before 18 February 1937 (the number suggests early February), SUR.

23. Wenger, *An Evaluation,* 15–16.

24. VDC Report 140, n.d., probably between 1 and 5 November, BR, Supplement II, folder 29.
25. Ibid.
26. VDC Report 164, 3 December 1937, BR, Supplement II, folder 31.
27. Wenger, *An Evaluation*, 18.
28. "List 35 Clinics in City for Free Syphilis Tests," *Chicago Tribune*, 5 September 1937, sec. A-8.
29. "Turns to Crime as Way to Get Syphilis Cure," *Chicago Tribune*, 10 October 1937, 24.
30. Wenger, *An Evaluation*, 17.
31. "Survey Shows Syphilis Drive Barely Begun," *Chicago Tribune*, 15 December 1937, 19.
32. "Proposes New City Venereal Disease Clinic," *Chicago Tribune*, 15 December 1937, 19.
33. "Offer Syphilis Tests to Relief and WPA Clients," *Chicago Tribune*, 4 February 1938, 6.
34. "Begin Free Tests Today in Drive to Find Syphilis," *Chicago Tribune*, 7 February 1938, 14.
35. Ibid.
36. "3,078 Take Free Syphilis Tests in Three Weeks," *Chicago Tribune*, 28 February 1938, sec. A-19.
37. "Clinics Follow Syphilis Trail, Find Carriers," *Chicago Tribune*, 7 March 1938, 11.
38. VDC Report 220, 17 February 1938, BR, Supplement II, folder 39.
39. VDC Report 245, 19 March 1938, BR, Supplement II, folder 43.
40. VDC Report 269, 19 April 1938, BR, Supplement II, folder 46.
41. VDC Report 85, 3 July 1937, SUR.
42. VDC Report 259, 5 April 1938, BR, Supplement II, folder 45.
43. "36 Permanent Syphilis Testing Offices to Open," *Chicago Tribune*, 4 April 1938, 10.
44. Wenger, *An Evaluation*, 29.
45. Ibid.
46. VDC Report 271, 21 April 1938, BR, Supplement II, folder 46. Quotations from the next two paragraphs refer to this same tour and are taken from this report.
47. Ibid.
48. VDC Report 19, 2 March 2, 1937, SUR.
49. VDC Report 274, 25 April 1938, BR, Supplement II, folder 46.
50. Ibid.
51. Duties, SUR.
52. VDC Report 269.
53. VDC Report 295, 20 May 1938, BR, Supplement II, folder 49.

54. A description of the Federal Theater, as well as the script of *Spirochete* and other plays from this series is printed in *Federal Theater Plays* (New York: Random House, 1938). A copy of the playbill for Chicago's production is included as an exhibit in Wenger's *An Evaluation*.

55. Ibid., 32.

56. Ibid., 29.

57. Ibid., 5.

58. VDC Report 317, 20 June 1938, BR, Supplement II, folder 52.

59. Ibid.

60. Jones, *Bad Blood*, 58–66.

61. "Syphilis Tests Increase 300 Pct. Over Last Year," *Chicago Tribune*, 10 May 1938, 11.

62. "In One Month 2,100 React to Syphilis Tests," *Chicago Tribune*, 28 June 1938, 10.

63. VDC Report 287, 11 May 1938, BR, Supplement II, folder 48.

64. VDC Report 309, 8 June 1938, BR, Supplement II, folder 50.

65. VDC Report 310, 10 June 1938, BR, Supplement II, folder 51.

66. VDC Report 324, 30 June 1938, BR, Supplement II, folder 52.

67. VDC Report 3, 30 July 1938, BR, Supplement II, folder 53.

68. Ibid.

69. VDC Report 323, 29 June 1938, BR, Supplement II, folder 52.

70. VDC Report 5, 5 August 1938, BR, Supplement II, folder 53; "Syphilis War Shows Advance on Two Fronts," *Chicago Tribune*, 5 August 1938, 10.

71. VDC Report 7, 6 August 1938, BR, Supplement II, folder 54.

72. VDC Report 120, 22 September 1937, BR, Supplement II, folder 24.

73. VDC Report 30, 9 September 1938, BR, Supplement II, folder 56.

74. VDC Report 13, 16 August, 1938, BR, Supplement II, folder 54.

75. Ben Reitman to Lawrence Linck, 6 March 1939, BR, Supplement II, folder 21.

Chapter 6: Net Results

1. "Alabama Joins War on Syphilis with Help for Doctors," *Chicago Tribune*, 6 August 1937, 13.

2. "7,000 Ft. Wayne Citizens Enroll in Syphilis War," *Chicago Tribune*, 28 October 1937; "Omaha Prepares for U.S. Aid in War on Syphilis" and "St. Louis Plans War on Syphilis," both in *Chicago Tribune*, 20 October 1937, 13.

3. "Michigan Plans Free Help in War on Syphilis," *Chicago Tribune*, 12 November 1937, 13; "New York Plans Free, Private Syphilis Tests," *Chicago Tribune*, 8 November 1937, 9. A brochure from New Jersey, detailing a program that includes testing, "Plain Facts: What Venereal Diseases Mean to New Jersey (Bureau

of Venereal Disease Control, New Jersey State Department of Health, cooperating with U.S. Public Health Service, November 1937), is contained in SUR.

4. VDC Report 72, n.d., between 21 May and 19 June 1937, SUR.

5. "All Income from Vast Fund Goes for Syphilis War," *Chicago Tribune*, 19 December 1937, sec. 1–20.

6. Wenger, *An Evaluation*, 4.

7. Chicago Syphilis Control Program, *Annual Report of the Chicago Syphilis Control Program, July 1, 1938–June 30, 1939* (Chicago, 1939), 38. A copy of this report is contained in BR, Supplement II, folder 17.

8. Wenger, *An Evaluation*, 4.

9. David C. Elliott to Thomas Parran, 8 March 1938, NA, Record Group 90, Box 28, folder 1850.

10. Wenger, *An Evaluation*, 6.

11. R. A. Vonderlehr to Thomas Parran, 11 December 1936, NA, Record Group 90, Box 27, folder 0425.

12. David C. Elliott to Thomas Parran, 8 March 1938.

13. R. A. Vonderlehr to David C. Elliott, 22 September 1938, NA, Record Group 90, Box 27, folder 0425/V.D.

14. Thomas B. Littlewood, *Horner of Illinois* (Chicago: Northwestern University Press, 1969), 83, 102–3.

15. Littlewood, *Horner*, 157–58.

16. Ibid., 175–77, quote from 177; Roger Biles, *Big City Boss in Depression and War: Mayor Edward J. Kelly of Chicago* (Dekalb: Northern Illinois University Press, 1984), 52–58.

17. Littlewood, *Horner*, p. 178.

18. VDC Report 115, 11 September 1937, BR, Supplement II, folder 24.

19. VDC Report 106, 25 August 1937, and VDC Report 114, 9 September 1937, both in BR, Supplement II, folder 24.

20. Articles about the negotiations around the new building appeared in the following issues of the *Chicago Tribune:* "New Laboratory Sought to Speed Syphilis Tests," 4 August 1937, 3; "School to Form Medical Center in Syphilis War," 10 August 1937, 11; "Examine School Proposed as GHQ in Syphilis War," 25 August 1937, 11; "Proposes New City Venereal Disease Clinic," 15 December 1937, 19; and "Aldermen Hear Plea for Money in Syphilis War," 16 December 1937, 18.

21. David C. Elliott to Thomas Parran, 12 August 1937, NA, Record Group 90, Box 28, folder 1850.

22. Undated speech by Lawrence J. Linck identified in BR, Supplement II, folder 22. The following description and quotations of the unit are all drawn from this talk.

23. David C. Elliott to Thomas Parran, 8 March 1938.

24. "Install Machine to Tabulate All Syphilis Cases," *Chicago Tribune*, 2 March 1938, 9.

25. VD Report 178, 21 December 1937, BR, Supplement II, folder 33.

26. Ibid.

27. Some early examples of his work in this area include a detailed list of suggested corrections to an unidentified twelve-page report ("Suggestions," SUR); a carefully explained report of "Correction in Chicago V.D. Bulletin, Volume 1, December 1937, No.6" (SUR—this bulletin was a regular publication of the Chicago Board of Health); and a letter from Lawrence Linck to Reitman acknowledging that he had sent Reitman certain sets of figures the inspector had requested in their meeting the day before (6 January 1938, SUR). Such documents run throughout Reitman's tenure in his position.

28. VD Report 219, 16 February 1938, BR, Supplement II, folder 34.

29. VDC Report 70, n.d., between 21 May and 19 June 1937, SUR.

30. VDC Report 97, 4 August 1937, BR, Supplement II, folder 23.

31. VD Report 123, 3 October 1937, BR, Supplement II, folder 26.

32. VDC Report 117, 15 September 1937, BR, Supplement II, folder 24.

33. VDC Report 245, 19 March 1938, BR, Supplement II, folder 44.

34. VDC Report 136, 27 October 1937, BR, Supplement II, folder 28.

35. "In One Month 2,100 React to Syphilis Tests," *Chicago Tribune*, 28 June 1938, 10.

36. "Uniform Remedy Is Next Goal in War on Syphilis," *Chicago Tribune*, 26 October 1937, 16.

37. VDC Report 135, n.d., probably between 23 and 27 October 1937, BR, Supplement II, folder 28.

38. Ibid.

39. Two collections of correspondence (NA, Record Group 90, Box 52, folder 0425 Gen. V.D. 1939 and Box 164, folder 0425 V.D. Gen. 1937) address this issue, which will be narrated fully in chapter 8.

40. VDC Report 133, n.d., probably between 20 and 23 October 1937, BR, Supplement II, folder 28.

41. VDC Report 201, 25 January 1938, BR, Supplement II, folder 36.

42. VDC Report 170, 10 December 1937, BR, Supplement II, folder 32.

43. VDC Reports 150, 17 November 1937, BR, Supplement II, folder 30; and 157, n.d., probably between 20 and 25 November 1937, BR, Supplement II, folder 31.

44. "Syphilis War Shows Advance on Two Fronts," *Chicago Tribune*, 5 August 1938, 10.

45. Wenger, *An Evaluation*, 37–38.

46. VDC Report 141, n.d., probably between 1 and 5 November 1937, BR, Supplement II, folder 29.

47. R. A. Vonderlehr to O. C. Wenger, 12 May 1938, NA, Record Group 90, Box 29, folder 0425.

48. Thomas Parran to O. C. Wenger, 6 May 1938, NA, Record Group 90, Box 29, folder 0425.

49. Thomas Parran to O. C. Wenger, 18 July 1941, NA, Record Group 90, Box 29, folder 0425.

50. Illinois Department of Public Health, *Annual Report of the Division of Venereal Disease, January 1, 1941, through December 31, 1941* (Springfield, 1941), 4.

Chapter 7: Privacy

1. "940 on West Side Apply for Free Syphilis Tests," *Chicago Tribune*, 22 May 1938, sec. 3W-5.

2. "Tribune Staff Asked to Take Syphilis Tests," *Chicago Tribune*, 15 March 1938, 5.

3. *Annual Report of the Surgeon General 1938*, 138.

4. "Concerns Hiring 15,000 Support Syphilis Tests," *Chicago Tribune*, 17 March 1938, 15; "Chicago Labor Backs Campaign against Syphilis," *Chicago Tribune*, 21 March 1938, 6.

5. "Dry Zero Staff of 152 Completes Syphilis Tests," *Chicago Tribune*, 9 April 1938, 5; "Safeguard Jobs in Syphilis War, Physicians Urge," *Chicago Tribune*, 9 June 1938, 14; "War on Syphilis in 2,800 Illinois Factories Urged," *Chicago Tribune*, 26 June 1938, sec. 1-6; "Laundry Owners Join City's War against Syphilis," *Chicago Tribune*, 6 August 1938, 3.

6. "War on Syphilis in 2,800 Illinois Factories Urged"; Wenger, *An Evaluation*, 30.

7. *Annual Report of the Surgeon General 1938*, 138.

8. R. A. Vonderlehr to David C. Elliott, 28 September 1937, NA, Record Group 90, Box 29, folder 0425.

9. Albert E. Russell to Thomas Parran, 2 April 1938, NA, Record Group 90, Box 29, folder 0425.

10. Ibid.

11. Wenger, *An Evaluation*, 24–25.

12. "Tells Industry Syphilis Tests Mean Economy," *Chicago Tribune*, 20 March 1938, sec. 1A-18.

13. "Surgeon General Calls upon Industry to Fight Syphilis," *Chicago Tribune*, 24 February 1938, 14; "Concerns Hiring 15,000 Support Syphilis Tests," *Chicago Tribune*, 17 March 1938, 15.

14. Wenger, *An Evaluation*, 25.

15. "Chicago Labor Backs Campaign against Syphilis," *Chicago Tribune*, 21 March 1938, 6.

16. Ibid., 25.

17. R. A. Vonderlehr to O. C. Wenger, 18 January 1939, NA, Record Group 90, Box 29, folder 0425.

18. "Laundry Owners Join Chicago's War against Syphilis."

19. Lida Usilton to David C. Elliott, 22 August 1939, NA, Record Group 90, Box 29, folder 0425. The correspondence between Shafer and Usilton that began in April 1939 and is contained throughout this box (and scattered through other parts of the Record Group) is interesting in itself because it illustrates the mutual support between two prominent women in male-dominated professions.

20. "Tribune Staff Asked to Take Syphilis Tests."

21. "Urges Insurance Men to Require Syphilis Tests," *Chicago Tribune*, 11 January 1938, 2; VDC Report 190, 12 January 1938, BR, Supplement II, folder 34.

22. "Widening War on the Venereal Plague," *Chicago Tribune*, 20 January 1938, 10.

23. VDC Report 196, 20 January 1938, BR, Supplement II, folder 35.

24. *Tribune* story cited in VDC Report 216, 14 February 1938, BR, Supplement II, folder 39.

25. VDC Report 225, 22 February 1938, BR, Supplement II, folder 40.

26. VDC Report 9, 9 August 1938, BR, Supplement II, folder 54.

27. David C. Elliott to Thomas Parran, monthly report for November 1937, NA, Record Group 90, Box 28, folder 1850.

28. Arthur Evans, "County Council of Legion Backs War on Syphilis," *Chicago Tribune*, 5 August 1937, 5.

29. In fact, in a letter from Vonderlehr to Wenger on 14 October 1937, Vonderlehr expressly stated that he hoped that the association would become "aggressive" in their educational program to younger people (NA, Record Group 90, Box 52, folder 0425). Shafer makes a similar comment in her remarks at "A Meeting Called by the Honorable Edward J. Kelly."

30. Thornton Smith, "Syphilis Peril to Youth Told by Dr. Parran," *Chicago Tribune*, 6 July 1937, 15; "Syphilis Survey to Be Conducted at 144 Colleges," *Chicago Tribune*, 31 December 1937, 3.

31. "Colleges to Join War to Stamp out Syphilis," *Chicago Tribune*, 24 September 1937, 10.

32. Chicago Syphilis Control Program, *Annual Report 1938–1939*, 48; BR, Supplement II, folder 18.

33. Chicago Syphilis Control Program, *Annual Report 1938–1939*, 48–49.

34. "N.U. Students Draft Syphilis Education Plan," *Chicago Tribune*, 27 March 1938, sec. 1A-5.

35. "350 Volunteer at N.U. to Take Syphilis Tests," *Chicago Tribune*, 31 March 1938, 14.

36. "Colleges Join in Campaign to Check Syphilis," *Chicago Tribune*, 20 January 1938, 7.

37. "Attacks Stand of Educators on Syphilis Menace," *Chicago Tribune*, 26 January 1938, 12.

38. "U. of Illinois Joins Student Paper's Anti-Syphilis Fight," *Chicago Tribune*,

23 October 1937, 3; "Champaign Dives Open; Police Fail to Make Raids," *Chicago Tribune*, 26 October 1937, 4; "2 Are Expelled from U. of I. in Cleanup of Vice," *Chicago Tribune*, 28 October 1937, 3. The later newspaper story gives the number of houses of prostitution as fifteen rather than the fourteen cited in the earlier story.

39. Chicago Syphilis Control Program, *Annual Report 1938–1939*, 26–27.

40. Wenger, *An Evaluation*, 22; "Launch Two New Drives against Syphilis Today," *Chicago Tribune*, 2 November 1937, 6. The procedure for gaining permission from parents is outlined in the Chicago Syphilis Control Program, *Annual Report 1939–1940*, 78, in BR, Supplement II, folder 18.

41. Chicago Syphilis Control Program, *Annual Report 1938–1939*, 26–27.

42. "Find 18 Syphilis Cases in Tests of 1,265 Pupils," *Chicago Tribune*, 8 January 1938, 13.

43. Wenger, *An Evaluation*, 22.

44. "1,100 Englewood Pupils to Take Syphilis Test," *Chicago Tribune*, 3 May 1938, 16.

45. "Plan Teachers' Meeting for War on Syphilis," *Chicago Tribune*, 29 October 1938, 13.

46. See also Wenger's *An Evaluation*, 22.

47. "Urges Education of Youth in War against Syphilis," *Chicago Tribune*, 18 May 1938, 3.

48. "Schools Asked to Give Pupils Syphilis Facts," *Chicago Tribune*, 31 December 1937, 3.

49. "Find 18 Syphilis Cases in Tests of 1,265 Pupils."

50. "Syphilis Tests Given 2,000 Aid Society Wards," *Chicago Tribune*, 11 July 1937, sec. 1A-5.

51. Ibid.; "Another Attack on Syphilis," *Chicago Tribune*, 14 July 1937, 8.

52. "Syphilis Tests Given Wards of Illinois Society," *Chicago Tribune*, 7 January 1938, 13.

53. "Social Agencies Join to Aid Child Victims of Venereal Disease," *Chicago Tribune*, 6 January 1938, 4.

54. Wenger, *An Evaluation*, 29–30.

Chapter 8: Privilege

1. R. A. Vonderlehr to O. C. Wenger, 14 July 1939, NA Record Group 90, Box 29, folder 0425.

2. Requests for federal money for these autopsies was a part of the annual reports of the surgeon general for fiscal years 1937–38; see *Annual Report of the Surgeon General 1936*, 122; *Annual Report of the Surgeon General 1937*, 116; and *Annual Report of the Surgeon General 1938*, 131.

3. Jones, *Bad Blood*, 27–28, 106.

4. Fee, "Sin versus Science," 125.

5. Jones, *Bad Blood*, 41, 43.

6. Ibid., 52–58.

7. Ibid., 55.

8. In addition to Jones's presentation of Nurse Eunice Rivers (*Bad Blood*), see also Darlene Clark Hine, *Black Women in White: Racial Conflict and Cooperation in the Nursing Profession, 1890–1950* (Bloomington: Indiana University Press, 1989), 154–56; and David Feldsuh, *Miss Evers' Boys*.

9. Personal communication from Susan Reverby, 16 November 1993.

10. Jones, *Bad Blood*, 57–58, 70.

11. R. A. Vonderlehr to O. C. Wenger, 14 October 1937, NA, Record Group 90, Box 52, folder 0425.

12. Jones, *Bad Blood*, 113–15.

13. R. A. Vonderlehr to Frank J. Jirka, 25 June 1936, NA, Record Group 90, Box 27, folder 0425.

14. Wenger, *An Evaluation*, 2.

15. See, for example, VDC Reports 60, n.d., between 3 March and 17 May 1937, SUR; and 76, n.d., between 21 May and 19 June 1937, SUR.

16. Wenger, *An Evaluation*, 22.

17. VDC Report 60.

18. VDC Report 97, 4 August 1937, BR, Supplement II, folder 23. There are two reports numbered 97, the other is dated 31 July.

19. VDC Report 192, 14 January 1938, BR, Supplement II, folder 35; VDC Report 269, 19 April 1938, BR, Supplement II, folder 46; VDC Report 281, 4 May 1938, BR, Supplement II, folder 47.

20. VDC Report 99, 10 August 1937, BR, Supplement II, folder 23.

21. VDC Report 197, 19 January 1938, BR, Supplement II, folder 36.

22. VDC Report 76.

23. VDC Report 126, 6 October 1937, BR, Supplement II, folder 26.

24. VDC Report 150, 17 November 1937, BR, Supplement II, folder 30.

25. Fee, "Sin versus Science," 126.

26. "Survey Shows Syphilis Drive Barely Begun," *Chicago Tribune*, 15 December 1937, 19.

27. Thomas Parran to John Lawlah, 20 December 1937, NA, Record Group 90, Box 164, folder 0425.

28. Thomas Parran to Col. McCormick, 23 December 1937, NA, Record Group 90, Box 164, folder 0425.

29. "Finds Syphilis Parallels Low Income Groups," *Chicago Tribune*, 28 December 1938, 16.

30. John Lawlah to Thomas Parran, 20 December 1937, NA, Record Group 90, Box 52, folder 0425.

31. S. W. Smith to Thomas C. Parran, 16 December 1937, NA, Record Group 90, Box 52, folder 0425.

32. Thomas Parran to S. W. Smith, 20 December 1937, NA, Record Group 90, Box 52, folder 0425, emphasis added.

33. Reported in VDC Report 184, 4 January 1938, BR, Supplement II, folder 33.

34. Chicago Syphilis Control Program, *Annual Report 1939–1940*, 90.

35. VDC Report 216, 14 February 1938, BR, Supplement II, folder 39; "Find Syphilis Cuts the Span of Life by 17 Percent," *Chicago Tribune*, 13 February 1938, sec. A-13.

36. Examples of Reitman's continued concern can be found in VDC Report 256, 30 [31?] March, 1938, BR, Supplement II, folder 45; VDC Report 4, 2 August 1938, BR, Supplement II, folder 53; VDC Report 24, 1 September 1938, BR, Supplement II, folder 56; and VDC Report 48, 3 October 1938, BR, Supplement II, folder 58.

37. VDC Report 256.

38. VDC Report 6, 5 August 1938, BR, Supplement II, folder 53 [VDC Report 5 also bears this date].

39. Bertha Shafer, in "Special Meeting Called by the Honorable Edward W. Kelley."

40. One of the classic articles presenting this concept is Carroll Smith-Rosenberg and Charles Rosenberg, "The Female Animal: Medical and Biological Views of Woman and Her Role in Nineteenth-Century America," *Journal of American History* 60 (September 1973): 332–56.

41. Quoted in *Chicago Tribune*, 8 October 1937, 19.

42. Editorial, "Widening War on the Venereal Plague," *Chicago Tribune*, 20 January 1938, 10; VDC Report 205, 31 January 1938, BR, Supplement II, folder 37.

43. Among Reitman's frequent references to such a law are those in VDC Report 147, n.d., probably between 9 and 17 November 1937, BR, Supplement II, folder 29; and VDC 285, 9 May 1938, BR, Supplement II, folder 48. The passage of the Illinois bill is noted Chicago Syphilis Control Program, *Annual Report 1938–1939*.

44. VDC Report 315, 16 June 1938, BR, Supplement II, folder 51.

45. VDC Report 76; VDC Report 315.

46. VDC Report 278, 29 April 1938, BR, Supplement II, folder 47.

47. VDC Report 263, 12 April 1938, BR, Supplement II, folder 46.

48. VDC Report 63, n.d., between 21 May and 19 June 1937, SUR.

49. VDC Report 56, n.d., between 3 March and 21 May 1937, SUR; VDC Report 81, n.d., between 21 and 27 June 1937, SUR.

50. VDC Report 78, n.d., between 21 May and 19 June 1937, SUR.

51. VDC Report 76; VDC Report 68, 21 May 1937, SUR.

52. For a discussion of race suicide, see Linda Gordon, *Woman's Body, Woman's Right* (Middlesex, England: Penguin, 1974), 137–40.

53. VDC Report 146, 10 November 1937, BR, Supplement II, folder 29 [a second VDC 146 is undated].

54. VDC Report 15, 19 August 1938, BR, Supplement II, folder 55.

55. Nine of the country's twenty-eight birth control clinics in 1929 were in Chicago, established by Hull-House resident and faculty at the University of Illinois, Dr. Rachelle Yarros. Patricia Spain Ward, "At the Eye of the Storm: Hull-House and the Chicago Birth Control Debate," presented at a symposium, "Hull-House and the People's Health," Chicago, 7 April 1990.

56. VDC Report 15.

57. "Syphilis Survey Shows Need for Care of Unborn," *Chicago Tribune*, 17 December 1937, 25.

58. VDC Report 285.

59. Chicago Syphilis Control Program, *Annual Report 1938–1939*, 34.

60. Ibid., 40.

61. VDC Report 78.

62. VDC Report 104, 18 August 1937, BR, Supplement II, folder 23.

63. VDC Report 158, 26 November 1937, BR, Supplement II, folder 31.

64. VDC 91, July 17, 1937, SUR.

65. VDC Report 99.

66. Marian [last name not given] to Ben Reitman, June 2, 1931, SUR.

67. VDC Report 138, 31 October 1937, BR, Supplement II, folder 28; see also VDC Report 81.

68. VDC Report 227, 24 February 1938, BR, Supplement II, folder 41.

69. VDC Report 99.

70. VDC Report 73, n.d., between 21 May and 19 June 1937, SUR.

71. VDC Report 299, 25 May 1938, BR, Supplement II, folder 49.

72. VDC Report 132, 20 October 1937, BR, Supplement II, folder 28.

73. VDC Report 291, 16 May 1938, BR, Supplement II, folder 48.

Chapter 9: Paranoia, Prudery, and Profit

1. Parran, *Shadow on the Land*, 57; Wenger, *An Evaluation*, 22.

2. Editorial, Irving S. Cutter, "The Syphilis Campaign and Fear," *Chicago Tribune*, 31 October 1989, 8.

3. VDC Report 198, 21 January 1938, BR, Supplement II, folder 36.

4. Brandt, *No Magic Bullet*, 50, 224n166.

5. VDC Report 79, 21 June 1937, SUR.

6. *Chicago Tribune*, 1 February 1939, 8.

7. Special Report 7, 7 February 1939, BR, Supplement II, folder 21.

8. *Chicago Tribune*, 14 June 1937, 15.

9. *Chicago Tribune*, 7 September 1937, 16.

10. VDC Report 113, 7 September 1937, BR, Supplement II, folder 24.

11. Reitman names a third clinic, the Civic Medical Center, in his reports but

gives no further information about it (VDC 62, n.d., between 3 March and 17 May 1937, SUR). It ran no advertisements in the *Tribune* during this period.

12. *Chicago Tribune*, 11 July 1937, 17.

13. *Chicago Tribune*, 1 February 1938, 10; 4 April 1938, 16.

14. All in the *Chicago Tribune*, 4 October 1937, 12; 5 July 1938, 10; 6 July 1937, 10; 3 January 1938, 10.

15. *Chicago Tribune*, 18 July 1938, 8.

16. *Chicago Tribune*, 1 June 1937, 14, 10.

17. *Chicago Tribune*, 7 March 1937, sec. 1–18.

18. *Chicago Tribune*, 12 September 1937, sec. 1A-18.

19. *Chicago Tribune*, 13 June 1937, sec. 1–12; 7 November 1937, sec. 1–14.

20. VDC 90, n.d., between 6 and 17 July 1937, SUR.

21. *Chicago Tribune*, 4 November 1934, sec. 1–25.

22. *Chicago Tribune*, 4 November 1934, sec. 1–18; 2 May 1935, 12.

23. VDC Reports 136, 27 October 1937, BR, Supplement II, folder 28; see also Report 185, 6 January 1938, BR, Supplement II, folder 33; and Report 191, 13 January 1938, BR, Supplement II, folder 34.

24. VDC Report 50, 5 October 1938, BR, Supplement II, folder 59.

25. VDC Report 127, 10 October 1937, BR, Supplement II, folder 27.

26. Special Report 7.

27. See, for example, VDC Report 87, 6 July 1937, SUR; and VDC Report 187, 8 January 1938 [also dated 7 January within the same report], BR, Supplement II, folder 34.

28. VDC Report 211, 8 February 1938, BR, Supplement II, folder 38.

29. VDC Report 105, 21 August 1937, BR, Supplement II, folder 24.

30. VDC Report 237, n.d., between 3 and 11 March 1938, BR, Supplement II, folder 43.

31. "Test Sanity of Al Capone," *Chicago Tribune*, 9 February 1938, 1.

32. "Illness Leaves Capone Calm, Says Warden," *Chicago Tribune*, 11 February 1938, 17; see also "Lawyer Plans Aid to Capone if Held Insane," *Chicago Tribune*, 10 February 1938, 4.

33. Brandt, *No Magic Bullet*, 155.

34. VDC Report 96, 30 July 1937, BR, Supplement II, folder 23. The folder contains two different reports bearing this number and date (or one report separated in the collection); this is the second of those reports.

35. Chicago Syphilis Control Program, *Annual Report 1938–1939*, 19.

36. "Schools Asked to Give Pupils Syphilis Facts," *Chicago Tribune*, 31 December 1937, 3.

37. "Funds to Teach Syphilis Danger Sought in Drive," *Chicago Tribune*, 6 March 1938, sec. 1–12.

38. VDC Report 154, 20 November 1937, BR, Supplement II, folder 30.

39. For more information on Dill Pickle Club, see Bruns, *Damndest Radical*, 230–45.

40. VDC Report 124, 29 September 1937, BR, Supplement II, folder 24, Reitman's ellipses.

41. VDC Report 66, 17 May 1937, SUR.

42. VDC Report 174, 15 December 1937, BR, Supplement II, folder 32.

43. VDC Report 314, 15 June 1938, BR, Supplement II, folder 51.

44. Brandt, *No Magic Bullet*, 114.

45. VDC Report 67, "Not Used," SUR [report followed by another 67, dated 20 May 1937].

46. *Annual Report of the Surgeon General 1937*, 118; *Annual Report of the Surgeon General 1938*, 135.

47. R. A. Vonderlehr to O. C. Wenger, 12 May 1938, NA, Record Group 90, Box 29, folder 0425.

48. R. A. Vonderlehr to O. C. Wenger, 26 February 1937, NA, Record Group 90, Box 29, folder 0425.

49. BR, Supplement II, folder 20. The sentence appears in a very brief, undated, undirected memo that has been copied in duplicate among Reitman's papers. All other items in this folder are dated from August through December 1938.

50. Walter Clarke to Ben Reitman, 24 February 1938, SUR.

51. Ben Reitman to Richard Finnigen, 14 February 1938, SUR.

52. Chicago Board of Health, *Division of Social Hygiene 1926*, 4–5.

53. VDC Report 162, 1 December 1937, BR, Supplement II, folder 31.

54. Ibid.

55. VDC Report 203, 28 January 1938, BR, Supplement II, folder 37.

56. R. A. Vonderlehr to O. C. Wenger, 12 May 1938.

57. VDC Report 128, n.d., between 10 and 17 October 1937, BR, Supplement II, folder 27.

58. VDC Report 64, n.d., between 3 March and 17 May 1937, SUR.

59. VDC Report 83, n.d., between 27 June and 3 July 1937, SUR.

60. VDC Report 42, 25 September 1938, BR, Supplement II, folder 58.

61. VDC Report 159, 27 November 1937, BR, Supplement II, folder 29.

62. VDC Report 161, 30 November or 1 December 1937, BR, Supplement II, folder 31.

63. *Webster's Dictionary* defines "grundyism" as "narrow prudish intolerant conventionality, especially as to the proprieties."

64. VDC Report 161.

65. VDC Report 198.

66. VDC Report 159.

67. Ben Reitman to R. A. Vonderlehr, 1 April 1937, SUR.

68. VDC Report 294, 19 May 1938, BR, Supplement II, folder 49.

69. VDC Report 67.

70. VDC Report 320, 25 June 1938, BR, Supplement II, folder 52.

71. VDC Report 223, 21 February 1938, BR, Supplement II, folder 40. See also VDC Reports 108, 2 September 1937, BR, Supplement II, folder 24; and 48, 3 October 1938, BR, Supplement II, folder 58.

72. See, for example, VDC Reports 119, 19 September 1937, BR, Supplement II, folder 24; and 5, 5 August 1938, BR, Supplement II, folder 53.

73. VDC Report 129, 17 October 1937, BR, Supplement II, folder 27.

74. VDC Report 115, 11 September 1938, BR, Supplement II, folder 24.

75. VDC Reports 192, 14 January 1938, BR, Supplement II, folder 35; and 56, 13 October 1938, BR, Supplement II, folder 59.

76. VDC Report 18, 3 March 1939, BR, Supplement II, folder 21.

77. VDC Report 282, 5 May 1938, BR, Supplement II, folder 47.

78. VDC Report 18.

79. VDC Report 41, 23 September 1938, BR, Supplement II, folder 58.

80. VDC Report 4, 2 August 1938, BR, Supplement II, folder 53.

Chapter 10: Last Straws

1. VDC Report 178, 21 December 1937, BR, Supplement II, folder 33.

2. R. A. Vonderlehr to O. C. Wenger, 7 August 1937, NA, Record Group 90, Box 27, folder 0243.

3. David C. Elliott to K. E. Miller, 19 October 1937, and K. E. Miller to David C. Elliott, 27 October 1937, both in NA, Record Group 90, Box 27, folder 0243.

4. NA, Record Group 90, Box 30, folder 0620.

5. Ben L. Reitman to W. A. Evans, 29 December 1938, BR, Supplement II-10, folder 94; Chicago Syphilis Control Program, *Annual Report 1938–1939*, title page.

6. O. C. Wenger to R. A. Vonderlehr, 29 March 1939, NA, Record Group 90, Box 30, folder 0620.

7. VDC Report 15, 19 August 1938, BR, Supplement II, folder 55.

8. "Offer Syphilis Tests to Relief and WPA Clients," *Chicago Tribune*, 4 February 1938, 6.

9. Biles, *Big City Boss in Depression and War*, 59, 78–79, 64–65.

10. VDC Report 114, 9 September 1937, BR, Supplement II, folder 24.

11. Clifford Blackburn, "J. Boyle, Hobo, Finds Chicago Relief a Cinch," *Chicago Tribune*, 13 January 1938, 1.

12. Seymour Korman, "Hobo Sam Hints He's a Fugitive but Gets Relief," *Chicago Tribune*, 15 January 1938, 4.

13. C. C. Applewhite to The Surgeon General, 3 January 1940, NA, Record Group 90, Box 28, folder 1616.

14. E. R. Coffey to C. C. Applewhite, 6 January 1940, NA, Record Group 90, Box 28, folder 1616.

15. Stories from the *Tribune* representative of the coverage of this topic include "Bare New Levy on Pay Rollers by Democrats," 5 March 1940, 1; "2 Pct Senators Fight New Curb on Pay Rollers," 6 March 1940, 2; and "Fail Raid on Clean Politics," 7 March 1940, 1.

16. The details of these events unfolded in the following articles in the *Chicago Tribune:* "Stelle Leaders See Widespread 20 Percent Slush Drive," 6 March 1940, 12; "Slush Fund's Foes Given Aid of Republicans," 8 March 1940, 1; "Reveal Smith Burned Files," 10 March 1940, 1, 2, 4; "Boss of 2 Percent Fund in on Secrets of Horner Machine," 10 March 1940, 1; "Slush Fund Suit Is Only Politics, Democrats Hold," 10 March 1940, sec. 1–5; "Smith-Horner Break Traced to Howe Case," 10 March 1940, sec. 1–3.

17. R. A. Vonderlehr to D. C. Elliott, 21 March 1940, NA, Record Group 90, Box 28, folder 1660.

18. Biles, *Big City Boss in Depression and War,* 56.

19. Ibid., 87.

20. "Investigation and Control of Venereal Diseases," Senate Report no. 1456, in *Senate Reports, 75th Congress, 2d and 3d Sessions, November 15, 1937–June 16, 1938* (Washington, D.C.: United States Government Printing Office, 1938), 1–2.

21. "Senate Group O.K.'s Big Fund to Fight Syphilis," *Chicago Tribune,* 16 February 1938, 13; "Approves Bill for State Aid in Syphilis War," *Chicago Tribune,* 24 February 1938, 14; "Senate Votes $12,000,000 to Fight Venereal Diseases," *Chicago Tribune,* 1 April 1938, 8; and "Approves Bill for U.S. Aid in War on Syphilis," *Chicago Tribune,* 21 April 1938, 11.

22. Copied in VDC Report 240, 14 March 1938, BR, Supplement II, folder 43.

23. VDC Report 301, 27 May 1938, BR, Supplement II, folder 49.

24. Editorial, "The New Federal Racket," *Chicago Tribune,* 18 February 1938, 10.

25. VDC Report 222, 19 February 1938, BR, Supplement II, folder 40.

26. "Chicago Center of Battle over Low Cost Clinic," *Chicago Tribune,* 1 August 1938, 2.

27. VDC Report 188, 8 January 1938, BR, Supplement II, folder 34.

28. Editorial, Morris Fishbein, "The Report of the Committee on the Costs of Medical Care," *Journal of the American Medical Association,* 10 December 1932, 2034–35, quoted by Patricia Spain Ward in "United States versus American Medical Association *et al.,* The Medical Antitrust Case of 1938–1943," *American Studies* 30 (Fall 1989): 123–54.

29. Quoted by Ann L. Wilson, "Development of the U.S. Federal Role in Children's Health Care: A Critical Appraisal," in *Children and Health Care: Moral and Social Issues,* ed. Loretta M. Kopelman and John C. Moskop (n.p.: Kluwer Academic Publishers, 1989), 27–65, quotation on 38. Subsequent discussion of medical opposition to child welfare measures is drawn from this article.

30. Quoted by Wilson, "Development of the U.S. Federal Role," 46–47.

31. Ibid., 53.

32. "Chicago Center of Battle over Low Cost Clinic."

33. Ward, "United States versus American Medical Association," 136, 143.

34. "Chicago Center of Battle over Low Cost Clinic."

35. VDC Report 4, 2 August 1938, BR, Supplement II, folder 53.

36. "Urges Education of Youth in War against Syphilis," *Chicago Tribune*, 18 May 1938, 3.

37. VDC 66, May 17, 1937, SUR.

38. "Urge Prevention Methods to Aid War on Syphilis," *Chicago Tribune*, 11 July 1938, 20. This "interview" was probably taken from an article by Schmidt in that month's *Chicago V.D. Bulletin*, which uses the exact wording as the *Tribune*'s quotation. The article is also quoted at length by Reitman in VDC Report 3, 30 July 1938, BR, Supplement II, folder 53.

39. VDC Report 3.

40. VDC Report 57, 14 October 1938, BR, Supplement II, folder 59.

41. Ibid.

42. Ben L. Reitman to R. A. Vonderlehr, November 1938, NA, Record Group 90, Box 164, Chicago Society for the Prevention of Venereal Disease.

43. Ben L. Reitman to A. C. Baxter, 5 November 1938, NA, Record Group 90, Box 164, Chicago Society for the Prevention of Venereal Disease.

44. The correspondence is sizable. From 15 March to 15 June 1937, for example, SUR contains three letters from Reitman to Vonderlehr (1 April, 14 April, 24 April); two letters from Reitman to Parran (15 March and 23 May); and four responses to Reitman from Vonderlehr (27 March, 7 April, 22 April, and 15 June).

45. Nels Anderson to Ben L. Reitman, 26 November 1938, NA, Record Group 90, Box 164, Chicago Society for the Prevention of Venereal Disease.

46. R. A. Vonderlehr to Ben L. Reitman, 2 December 1938, NA, Record Group 90, Box 164, Chicago Society for the Prevention of Venereal Disease.

47. Ben Reitman to Lawrence Linck, 6 January 1939, BR, Supplement II, folder 21; the identification of Leo Reitman's business appears in a letter from Reitman to William A. Evans, 3 June 1939, NA, Record Group 90, Box 164, Chicago Society for the Prevention of Venereal Disease.

48. "Bundesen to Go Before Jurors in Milk Inquiry," *Chicago Tribune*, 6 October 1938, 4; "Before Jury," *Chicago Tribune*, 8 October 1938, 8.

49. Chicago Syphilis Control Program, *Annual Report 1939–1940*, 130.

50. Ibid., 14–15.

51. Ibid., 33, 44.

52. Ibid., 92.

53. Ibid., 92–93.

54. Ibid., 93.

55. Ibid.

56. Ibid., 94.

57. Ibid., 95, emphasis added.

58. "New Deal Wrecked by G.O.P.," *Chicago Tribune*, 9 November 1938, 1.

59. Ben Reitman to R. L. Kahn, 4 October 1941, BR, Supplement II, folder 79.

Chapter 11: Aftermath

1. "Friends Eulogize Dr. Ben," *Chicago Tribune*, 20 November 1942. This and other eulogies were collected in a scrapbook located in SUR; pages of the clippings are not noted. Some of those other articles are: "Reitman Friends to Hear Eulogy by 'Yellow Kid," *Chicago Daily News*, 21 November 1942; "Dr. Reitman Estate $450, Probate Shows," *Chicago Sun*, 21 November 1942; "Yellow Kid Weil Pays Tribute to Reitman Tonight," *Chicago Tribune* 21 November 1942; and "Elite of W. Madison St. Pack Reitman's 'Send Off'," *Chicago Sun*, 20 November 1942.

2. Lawrence J. Linck to Ben Reitman, 5 October 1939, BR, Supplement II, folder 74.

3. Wenger, *An Evaluation*, 28; Chicago Syphilis Control Program, *Annual Report 1939–1940*, 10–13.

4. "Syphilis Tests Increase 300 Pct. over Last Year," *Chicago Tribune*, 10 May 1938, 11.

5. Chicago Syphilis Control Program, *Annual Report 1938–1939*, 7.

6. Chicago Syphilis Control Program, *Annual Report 1939–1940*, 10–13.

7. Ibid., 14–15.

8. Ibid., 10–13.

9. Ibid., 16.

10. Lida Usilton to O. C. Wenger, 19 July 1940, NA, Record Group 90, Box 27, folder 0425.

11. Chicago Syphilis Control Program, *Annual Report 1939–1940*, 52–53, 80, 60, 100, 45, 66.

12. Chicago Syphilis Control Program, *Annual Report 1938–1939*, 14.

13. Chicago Syphilis Control Program, *Annual Report 1939–1940*, 33.

14. Ibid., 33, 44–45.

15. Ibid., 23.

16. Ibid., 23–24.

17. See, for example, Gordon, *Woman's Body, Woman's Right;* and James Reed, *From Private Vice to Public Virtue: The Birth Control Movement and Americal Society since 1830* (New York: Basic Books, 1978).

18. Chicago Syphilis Control Program, *Annual Report 1938–1939*, 5, 19–21.

19. Chicago Syphilis Control Program, *Annual Report 1939–1940*, 121.

20. Jay Cassel, *The Secret Plague: Venereal Disease in Canada 1838–1939* (Toronto: University of Toronto Press, 1987).

21. Cassel, *The Secret Plague*, 145.

22. Ibid., 189.

23. Ibid., 176–78, 186–87, 189–91.

24. Ibid., 206–7, 214.

25. Ibid., 206–7, 214, 220.

26. Ibid., 219.

27. Ibid., 198–203.

28. Ibid., 205.

29. Ibid., 245.

30. Ibid., 221.

31. Ibid., 107; see also chapter 1.

32. Ibid., 120.

33. Ibid., 10; Brandt, *No Magic Bullet*, 161, 163.

34. O. C. Wenger, "The Need for Quarantine Facilities in the Chicago Syphilis Control Program," 19 July 1941, NA, Record Group 90, Box 29, folder 0425.

35. Herman Soloway, *Annual Report, Division of Venereal Diseases, Illinois Department of Public Health, January 1, 1941, through December 31, 1941* (Springfield, 1941), 2, NA, Record Group 90, Box 28, folder 1850; *Annual Report, Cook County Public Health Unit, July 1, 1941 through June 30, 1942*, NA, Record Group 90, Box 31, Cook County folder.

36. Memo from Assistant Surgeon General R. A. Vonderlehr to The Surgeon General, 2 July 1942, NA, Record Group 90, Box 27, folder 0425.

37. Wenger, "The Need for Quarantine facilities in the Chicago Syphilis Control Program."

38. Soloway, *Annual Report Division of Venereal Disease, Illinois Department of Public Health 1940–1941.*

39. Ibid.

40. Thomas Parran to Dr. I. V. Sollins, 31 July 1942, NA, Record Group 90, Box 27, folder 0425.

41. "Hotel Is Closed as a Venereal Disease Menace," *Chicago Tribune*, 24 June 1942, NA, Record Group 90, Box 0425.

42. Illinois Department of Public Health, *Annual Report, January 1 through December 31, 1941;* Memo from Passed Assistant Surgeon George E. Parkhurst to The Surgeon General, July 8, 1944. Both documents are located in NA, Record Group 90, Box 27, folder 0425.

43. George E. Parkhurst to The Surgeon General, 8 July 1944.

44. Howard W. Ennes, Jr., to A. J. Aselmeyer, 23 August 1940, NA, Record Group 90, Box 52, folder 0425.

45. A. J. Aselmeyer to O. C. Wenger, 8 October 1940, NA, Record Group 90, Box 52, folder 0425.

46. Ibid.

47. I. V. Sollins to Dr. Thomas Parran, 8 July 1942, NA, Record Group 90, Box 27, folder 0425.

48. NA, Record Group 90, Box 27, folder 0425.

49. "Charges Navy Sponsors Jap Vice Resorts," *Chicago Tribune*, 27 November 1945, 9.

50. Thomas Devine, Director, Social Protection Division, to Howard F. Feast, Regional Social Protection Representative, Chicago, 24 May 1945, NA, Record Group 215, Box 6, folder 849.

51. The correspondence tracing the rise and demise of the early treatment centers is contained in correspondence among Bundesen, Parran, and Vonderlehr in several locations within the National Archives, primarily Record Group 90, Box 27, folders 0135 and 0425.

52. Brandt, *No Magic Bullet*, 170.

Epilogue: Between Syphilis and AIDS

1. Frank E. James, "Move to Repeal Illinois AIDS Law Signals Less Strict Trend in States," *Wall Street Journal*, 9 January 1989, sec. B-4.

2. Ibid., brackets in original quote.

3. Two early examples are Allan M. Brandt, "AIDS in Historical Perspective: Four Lessons from the History of Sexually Transmitted Diseases," *American Journal of Public Health* 78 (April 1988): 367–71; and Gregg S. Meyer, "Criminal Punishment for the Transmission of Sexually Transmitted Diseases: Lessons from Syphilis," *Bulletin of the History of Medicine* 65 (1991): 549–64.

4. The words *war* and *plague* and language relating to those terms has peppered the documents cited from the time of the Chicago Syphilis Control Program. For similar discussions about similar discourse around AIDS and HIV, see Sander Gilman, "Plague in Germany, 1939/1989: Cultural Images of Race, Space, and Disease," 54–82, and Michael S. Sherry, "The Language of War in AIDS Discourse," 39–53, both in *Writing AIDS: Gay Literature, Language, and Analysis*, ed. Timothy F. Murphy and Suzanne Poirier (New York: Columbia University Press, 1993); and Paula A. Treichler, "AIDS, Homophobia, and Biomedical Discourse: An Epidemic of Signification," in *AIDS, Cultural Analysis/Cultural Activism*, ed. Douglas Crimp (Cambridge: The MIT Press, 1987), 30–70.

5. One analysis that takes up the question of gay male "promiscuity" is found in Mary Catherine Bates and William Goldsby's *Thinking AIDS: The Social Response to the Biological Threat* (Reading: Addison Wesley, 1988), 42–44.

6. Treichler, "AIDS, Homophobia, and Biomedical Discourse," 43; Timothy E. Cook and David C. Colby, "The Mass-Mediated Epidemic: The Politics of AIDS on the Nightly Network News," in *AIDS: The Making of a Chronic Disease*, ed. Elizabeth Fee and Daniel M. Fox (Berkeley: University of California Press, 1992), 49–83.

7. Larry Kramer, *Reports from the Holocaust: The Making of an AIDS Activist* (New York: St Martin's Press, 1989), 13.

8. C. Everett Koop, *Koop: The Memoirs of America's Family Doctor* (New York: Random House, 1991), 238.

9. Ronald Bayer, "Entering the Second Decade: The Politics of Prevention, the Politics of Neglect," in *AIDS: The Making of a Chronic Disease*, ed. Fee and Fox, 207–26, esp. 222.

10. Koop, *Koop*, 127.

11. Ibid., 6, 149, 195–96, 223.

12. Ibid., 77.

13. Ibid., 102.

14. Ibid., 179–80.

15. Ibid., 186.

16. Ibid., 59.

17. A discussion of this event took place in a workshop on activism conducted by Tim Miller and Jeanne Kracher, "Chicago Activists—ACT UP" during the conference "AIDS: Images, Actions, and Analysis," School of the Art Institute of Chicago, Chicago, Illinois, 1–2 December 1992.

18. Robert A. Padgug and Gerald M. Oppenheimer, "Riding the Tiger: AIDS and the Gay Community," in *AIDS: The Making of a Chronic Disease*, ed. Fee and Fox, 258, 269.

19. Ibid.

20. Don C. Des Jarlais, Samuel R. Friedman, and Jo L. Sotheran, "The First City: HIV among Drug Users in New York City," in *AIDS: The Making of a Chronic Disease*, ed. Fee and Fox, 288; Miller and Kracher, "Chicago Activists—ACT UP."

21. Kramer, *Reports from the Holocaust*, 102.

22. Des Jarlais, Friedman, and Sotheran, "The First City," pp. 282–86; Richard Elovich, "Addicts, AIDS, and Notions of Community Building," Steve Wakefield, "'We're All in This Together'—Who Says?—The Psychological Cost of Being Invisible: An Analysis of the Effect of African-American Images of HIV/AIDS Education and Services," and Robert Vázquez, "Marginalization 'N Shit," all presented at the Chicago conference "AIDS: Images, Actions, and Analysis"; Risa Denenberg, "What the Numbers Mean," 1–4, and Ruth Rodriguez, "We Have the Expertise, We Need the Resources," 99–101, both in *Women, AIDS and Activism*, by the ACT UP/New York Women and AIDS Book Group (Boston: South End Press, 1990).

23. Padgug and Oppenheimer, "Riding the Tiger," 268. The difficulty of keeping cultural sensitivity an integral part of all discussions and planning is addressed throughout the essays in *Women, AIDS, and Activism* by the ACT UP/New York Women and AIDS Book Group. It was also addressed by the following speakers at the Chicago conference "AIDS: Images, Actions, and Analysis": Wakefield, "'We're All in This'"; Elovich, "Addicts, AIDS, and Notions"; Vázquez, "Marginalization"; and Robert Atkins, "Visual AIDS and Day without Art: Or, How to Have Art Events in an Epidemic That Might Reach and Teach Diverse Audiences."

24. Koop, *Koop*, 97.

25. Kramer, *Reports from the Holocaust*, xvii.

26. Douglas Crimp with Adam Rolston, *AIDS/demo/graphics* (Seattle: Bay Press, 1990), 32–33.

27. Koop, *Koop*, 238.

28. Cook and Colby, "The Mass-Mediated Epidemic"; Crimp and Rolston, *AIDS/demo/graphics*.

29. Paula A. Treichler, "Transgression and Containment at Home and Abroad: Representations of Global AIDS," presented at the Chicago conference "AIDS: Images, Actions, Analysis"; *Life Magazine*, July 1985.

30. Douglas Crimp, "How to Have Promiscuity in an Epidemic," in *AIDS: Cultural Analysis/Cultural Activism*, ed. Crimp, 237–371, citation on 263.

31. Ibid., 266.

32. Koop, *Koop*, 238; Cook and Colby, "Mass-Mediated Epidemic," 113.

33. Cook and Colby, "Mass-Mediated Epidemic," 113–14.

34. Kramer, *Reports from the Holocaust*, 33.

35. Ibid., 14.

36. Ibid., 187.

37. Martha Gever, "Pictures of Sickness: Stuart Marshall's *Bright Eyes*," in *AIDS: Cultural Analysis/Cultural Activism*, ed. Crimp, 108–26, quotation on 121.

38. Koop, *Koop*, 211.

39. Kramer, *Reports from the Holocaust*, xvii.

40. Paula A. Treichler, "AIDS, Gender, and Biomedical Discourse: Current Contests for Meaning," in *AIDS: The Making of a Chronic Disease*, ed. Fee and Fox, 190–266; Zoe Leonard and Polly Thistlethwaite, "Prostitution and HIV Infection," in *Women, AIDS, and Activism*, by the ACT UP/New York Women and AIDS Book Group, 177–86.

41. "*Pneumocystis* Pneumonia—Los Angeles" and "Kaposi's Sarcoma and *Pneumocystis* Pneumonia among Homosexual Men—New York City and California," both in *Morbidity and Mortality Weekly Report* 30 (1981): 250–52, 305–8. Much has been written on this equation; see, for example, James W. Jones, "Refusing the Name: The Absence of AIDS in Recent American Gay Male Fiction," in *Writing AIDS*, ed. Murphy and Poirier, 225–43.

42. Gerald M. Oppenheimer, "In the Eye of the Storm: The Epidemiological Construction of AIDS," in *AIDS: The Burdens of History*, ed. Fee and Fox, 262–300; and Gerald M. Oppenheimer, "Causes, Cases, and Cohorts: The Role of Epidemiology in the Historical Construction of AIDS," in *AIDS: The Making of a Chronic Disease*, ed. Fee and Fox, 49–83.

43. Koop, *Koop*, 224.

44. Ibid., 226.

45. Ibid., 195.

46. Ibid., 85.

47. Ibid., 209.

48. Parran, *Shadow on the Land*, 222–23.

49. Koop, *Koop*, 237.

50. Kramer, *Reports from the Holocaust*, 178.

51. Larry Kramer, *Faggots* (New York: Random House, 1978).

52. Joseph Cady, "Immersive and Counterimmersive Writing about AIDS: The Achievement of Paul Monette's *Love Alone*," 244–64, and James Miller, "Dante on Fire Island: Reinventing Heaven in the AIDS Elegy," 265–305, both in *Writing AIDS*, ed. Murphy and Poirier.

53. Padgug and Oppenheimer, "Riding the Tiger," 261.

54. Various aspects of this debate are taken up by Lee Edelman, "The Mirror and the Tank: 'AIDS,' Subjectivity, and the Rhetoric of Activism," 9–3, John M. Clum: " 'And Once I Had It All': AIDS Narratives and Memories of an American Dream," 200–224, and Miller, "Dante on Fire Island: Reinventing Heaven in the AIDS Elegy," all in *Writing AIDS*, ed. Murphy and Poirier; see also Crimp, "How to Have Promiscuity."

55. Risa Denenberg, "What the Numbers Mean," Ruth Rodriguez, "We Have the Expertise, We Need the Resources," both in *Women, AIDS, and Activism*, by the ACT UP/New York Women and AIDS Book Group; Daniel M. Fox, "The Politics of HIV Infection: 1989–1990 as Years of Change," in *AIDS: The Making of a Chronic Disease*, ed. Fee and Fox, 125–43; Wakefield, " 'We're All in This Together' "; Vázquez, "Marginalization."

56. Douglas Crimp, "Accommodating Magic," presented at the Chicago conference "AIDS: Images, Actions, and Analysis"; Harlon L. Dalton, "AIDS in Blackface," *Daedalus* 118 (1989): 205–27; Phillip Brian Harper, "Eloquence and Epitaph: Black Nationalism and the Homophobic Impulse in Responses to the Death of Max Robinson," in *Writing AIDS*, ed. Murphy and Poirier, 117–39.

57. Crimp, "Accommodating Magic," Harper, "Eloquence and Epitaph," and Timothy F. Murphy, "Celebrity AIDS," in *Ethics in an Epidemic: AIDS, Morality, and Culture* (Berkeley: University of California Press, 1994), 69–81.

58. Des Jarlais, Friedman, and Sotheran, "The First City," 282–86; Richard Elovich, "Addicts, AIDS, and Notions of Community Building," presented at the Chicago conference "AIDS: Images, Actions, and Analysis."

59. For example, Timothy F. Murphy, "The Angry Death of Kimberly Bergalis," *Proteus* 9 (1992): 3–10; "Testimony," in *Writing AIDS*, ed. Murphy and Poirier, 306–20; Cindy Ruskin, Matt Herron, and Deborah Zemke, *The Quilt: Stories from the Names Project* (New York: Pocket Books, 1988); *AIDS: The Women*, ed. Ines Rieder and Patricia Ruppelt (San Francisco: Cleis Press, 1988); and *Positive Women: Voices of Women Living with AIDS*, ed. Andrea Rudd and Darien Taylor (Toronto: Second Story Press, 1992).

60. Treichler, "AIDS, Homophobia, and Biomedical Discourse," 33.

61. At least, perhaps, until recently, as the appearance of new, penicillin-resistent strains of syphilis has prompted a renewed flurry of medical and public health concern. See Kitry Krause, "On the Streets Where Syphilis Lives: Tracking the Epidemic One Name at a Time," *Chicago Reader*, 13 November 1992.

Index

SUZANNE POIRIER is associate professor of literature and medical education at the University of Illinois at Chicago. She is on the faculty of the Medical Humanities Program, an interdisciplinary faculty whose teaching and research address the cultural and personal contexts of medical practice. She is editor of the journal *Literature and Medicine* and coeditor, with her colleague Timothy F. Murphy, of *Writing AIDS: Gay Literature, Language, and Analysis.*

Printed and bound by CPI Group (UK) Ltd, Croydon, CR0 4YY

16/04/2025

14658439-0002